BEYOND THE FABRIC OF EXISTENCE

By Wayne Thompson

I would like to thank God and the many people over the years that have inspired me to continue my work.

A special thank you to my friend Cody Wilson for the use of his painting for the Front Cover and illustration on the Back Cover.

INDEX

Holographic Universe   7

Vedic Physics and the Fifth Element   51

The Vedic Metric and the Lotus Flower Measuring system   84

Intelligent Design   123

Ancient Astronauts & Suppressed Information   174

Trust, the System of Deception   338

Conclusion   369

Preface

There have been several scientific books and lecture papers written on the subject of our holographic universe but none have gone far enough as to expand peoples thinking and explain the true nature of reality. Music is a natural consequence of the pure mathematics within nature. Music is a true universal language as Music is vibrational physics and mathematics that is a language understood by the human mind. The silent music of the universe or Aether Physics from the RG Veda is the only ONE science that explains the true perfection of creation and our connection to the holographic universe.

Quantum Metrics are from the RG Veda: Quantum Physicist already knowing the answer as they have taken it the RG Veda then creates complicated elongated mathematical equations to derive at their Metric, which they name after themselves. I explain how to calculate all 90 metrics contained in RG Veda using a dividend and divisor and how to apply this system of harmony to devices you can manufacture such as electric motors. I would not dare name any of the yet "undiscovered" Metrics after myself, as no man should claim Gods work as his own.

Although I have examples of the RG Vedas and other sources mentioning the Vedic Meter no one to my knowledge as given a full interpretation of them and what they relate to as I have done. I have deciphered and attempted to simplify one of the most ancient of mysteries and show how to apply it.

My intention in releasing this information is to enlighten humanity as to assist in the rebuilding of the foundations of science for the advancement of all. We all must aspire to a brighter future and not allow this information to remain the industrial secret of occult societies.

These societies have handicapped humanity for long enough and it is time to enter into the light from the darkness and advance our civilization. The zenith is the point in the sky or celestial sphere directly above an observer. God, sees all life in all dimensions and knows all of us, we should all strive for Krsna Consciousness and free ourselves from the illusion of our material world. When there is harmony between the mind, heart and resolution then nothing is impossible

Holographic Universe

The holy trinity, the Trimūrti or triple form. This consisted of Brahmā the creator, Vishnu the preserver, and Śiva the destroyer. 'The Trinity of Brahmā, Vishnu, and Śiva are the threefold manifestations of the one Supreme, Krsna.'

The lotus flower measuring system, the measuring system of the gods is as old as creation itself and you can see all three are standing on lotus flowers from which this system is derived. There are clearly different design principles contained within nature's glorious diversity and I show different examples of as many as I can as to help encourage people to what to know more.

Living in a Hologram

Your eyes just interpret and receive information like antennas receiving information emitted by the machine. There is really nothing in front of you all it is played out in your head like the projected images on the television.

We are bioelectric emitter receivers, which a machine records our every action. The Human mind is a glorious thing but it does deceive us and inhibits or perhaps shields us from the seeing reality as the illusion it is. Holograms contain all the information needed to reconstruct the entire image. They contain many dimensions of information in far less space, like a compressed file. They hold that information in a network of interacting frequencies. Like shining a coherent light (reference beam) or laser through the blurred overlapping waves of a two-dimensional hologram, you can create a virtual image of a three-dimensional figure.

Our brains mathematically construct a solid reality by interpreting frequencies from another dimension. This information realm of meaningful, patterned, primary reality transcends time and space. Thus, the brain is an embedded hologram, interpreting a holographic Universe. All existence consists of embedded holograms within holograms, and their interrelatedness somehow gives rise to our existence and sensory images.

There is this excellent documentary called the **Holographic Universe – Beyond Matter:** It explains the Physical Universe as being an illusion and your senses create electrical signals in the back of our brains. This means that we have absolutely no idea of what exists outside of our mind; we cannot see it, only the electrical impulses of something transmitted from outside of ourselves. When we look at a mountain, for example, we perceive it as on object separated from ourselves, when in fact it exists as a hologram of electrical impulses inside of us.

Then, if we expand on this, what about the brain? If all the rest is a hologram, so is the brain. Who is it then that "sees"? There is only one explanation - the soul or the spirit! This is now scientific fact. Philosophers since the Greeks have speculated about the "ghost" in the machine, the 'little man inside the little man' and so on. Where is the I – the entity that uses the brain? Who does the actual knowing? Or, as Saint Francis of Assisi who died October 3, 1226 once put it, "What we are looking for is what is looking."

The Source of this hologram is a machine and they have known that for thousands of years. Norbert Wiener, same say the founder of Cybernetics stated many years ago "information is information, not matter and not energy". He also concludes "that beyond the wave is a vortex of information produced by an external artificial source" this artificial source forms our existence. There has been much written about the electric contact universe and our holographic universe since about 1946 and it has become an ideology familiar to many. We do live in holographic universe and like all projected holograms there must be a source from which it comes.

With this in memory, a chilling question comes to mind: If this is true, WHO OR WHAT IS MANIPULATING US FROM OUTSIDE TO PERCEIVE WHAT WE PERCEIVE ON THE INSIDE?

Are we actually (spiritually) hooked up to something like super computers, being totally enslaved by the illusion within our own minds, to think that the physical universe is real, and are we in the mercy of 'something' we can never find as we are connected to a 'super computer'? We are just pawns in the game of life, as they say the whole world is a stage. How can we break free, and what happens if we do? The above question comes to mind somewhere in our thought process, but there could be another explanation. Perhaps we are here in this hologram to learn lessons, in order to evolve into higher planes of consciousness.

Perhaps we were put here, with or without our consent, to learn how to make right choices. This means we were sent here with a purpose not to be interfered with; we were sent here with free will. It is up to us to what the outcome of our lives will be. Will we learn from our experiences or not? Are we learning enough to gain a better understanding to achieve the next level, or do we need to reincarnate in a future life to learn more?

The Spirit or Soul is certainly a complicated thing on this level, and I think that on this level of reality, which we call the 3rd Dimension, it is possible to get all questions answered but we must try no matter how difficult the task. I believe it is ESSENTIAL that we at least try to be Krsna conscious and maintain our mind as much as possible on a spiritual level while we are here.

The more we evolve our consciousness here, the easier it will be for us once we leave or bodies/vessels in a stage we know as "death" - another illusion as how can you die if you are not really alive, death is only the beginning. We need to have faith in the Supreme Being to guide our Spirit/Soul to a higher comprehension to avoid the entrapment of the illusion as not to repeat the mistakes of the past. I do not think anybody who watches this film will be unaffected by its contents. It will forever change your way of perceiving reality!

This is the statement from the beginning of the movie:

The subject of this film you are about to watch reveals a crucial secret of your life. You should watch it very attentively for it concerns a subject that is liable to make fundamental changes in your outlook on the material world. The content of this film is not just a different approach, or a philosophical thought: It is a fact that is also proven by science today. Still, there is this ultimate question:

If we are pure spirit, with no location in time and space, and all that exists outside spirit is THOUGHT, who created SPIRIT? Why is there LIFE at all? The only explanation I can find is that there is a GOD, which probably is pure energy. Personally, I believe there is a Creator, a loving force, who cares for our well-being and us.

I think we are all here on Earth to learn, and our bodies are our vessels. However, I think there is some truth in most religions and that we have to embark on our own journey of discovery. Ultimately, the truth is within ourselves, and this is where we will find the answers. Others before us have experienced similar dilemmas on their journey towards the truth but we must find our own paths.

We are all on a spiritual journey, and I think it is important that we explore life because we all know the truth is 'deep inside', we need to follow our intuition.

This film ends with suggesting that the Supreme Being (God) is Allah and when we die, we leave the holographic prison of a material reality and go to Him. Krishna, who is known to Muslims as Allah, or to the Christians as the Almighty father, or as Jehovah or Yahweh too many, is the most complete and attractive understanding of God. Krishna means 'All attractive'. A person is attractive if He possesses unlimited beauty, knowledge, strength, fame, riches or renunciation. Krishna alone possesses not only all the six opulence's but He has them in unlimited abundance.

We learn in the Bhagavad-Gita that total devotion to Krishna will allow man to achieve eternal life and never know death again. In the New Testament, we learn that Christ is the only way. 'That whosoever believeth in Christ should not perish, but have everlasting life.'

Logically thinking what does this tell you about this paradox?"
Here are two different religions revealing to us the creator of all things who, is also the way to achieve eternal life. Either one is a fraud or both are the same holy manifestation recorded at different times from two different cultural and historical viewpoints. The logic is clear and simple.

Krishna is the Christ of Christianity. Christ comes from the Greek word Christos, and Christos is the Greek version of the word Krishna. When an Indian person calls on Krishna, he often says "Krista." Krishna is a Sanskrit word meaning the object of attraction." So when we address God as "Christ," "Krista," or "Krishna," we indicate the same all-attractive Supreme Personality of Godhead.

They have found proof that Krishna existed! They found that Vedic texts are authentic! Krishna's lost city has also been found and at same place as told by Hindu texts like Mahabharata. They also have proof that Buddha existed, is there any proof that Christ ever existed? The only examples are the bible and gospels that are copies of Hindu texts.

 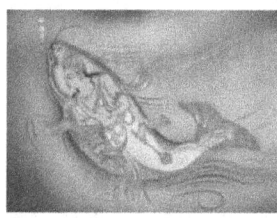 

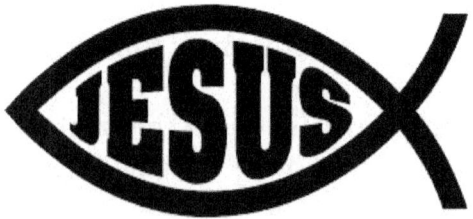 Krishna is Christ

The cosmic Christ. From the ManuScript10-ETIGammaNeT
""Cosmic Christ is the heart of all creation, and God which is All-Oneness, which is God. Ultimately cosmic Christ is God and all things are God, all this are One, and that is what we term All-Oneness.

"There is only One great unity, One great Being, that has divided Itself, in Its stillness, in Its Beingness, to be many different beings, to be many different universes, and many different creations, in which It can sub-divide Itself again, into many different galaxies, into many different solar systems; into many different planets; into many different beings in those planets; into many different molecules, atoms, sub atoms. Until you come to a stage, within subatomic particles, where everything is God and is All-Oneness. "For as you merge into Christ there is no size. For you become all things, you become all sizes, great and small. You can identify with every single thing that exists within this universe and other universes, within all dimensions — of which there are infinite varieties." The Rg Veda not only appears to be the root of much of the parables of Christ, laid down in the New Testaments 27 books, but its riches go too far greater depths in the treasury of its Sophia and philosophy. Coupled to this, Christ described in John as "the Word", is also described as the Agency through which "all things were made." And thereby, it would be natural to find the embroidery of the "Words" handy work in history. As the Book of Revelation ascribes the Lord as saying: "I Am the Alpha and Omega." The concept of the Alpha and Omega stems from the Rg Veda that describes the Alpha as Ruta, the imperishable root universal Order (the prefix, Ovoid, Oval as the All, overtly overall); and Omega as Satya, the pansperm. To me it appears that Christ (KRSNA) consciousness is a way to conduct yourself in the manner of Gods thinking.

Structure of the Holographic Brain

Your eyes interpret information and your mind converts this information into holographic images and records them, this is what you call memories. The below images I saved some time ago from a website that now no longer exists, Therefore, I am unable to give credit to whoever created them.

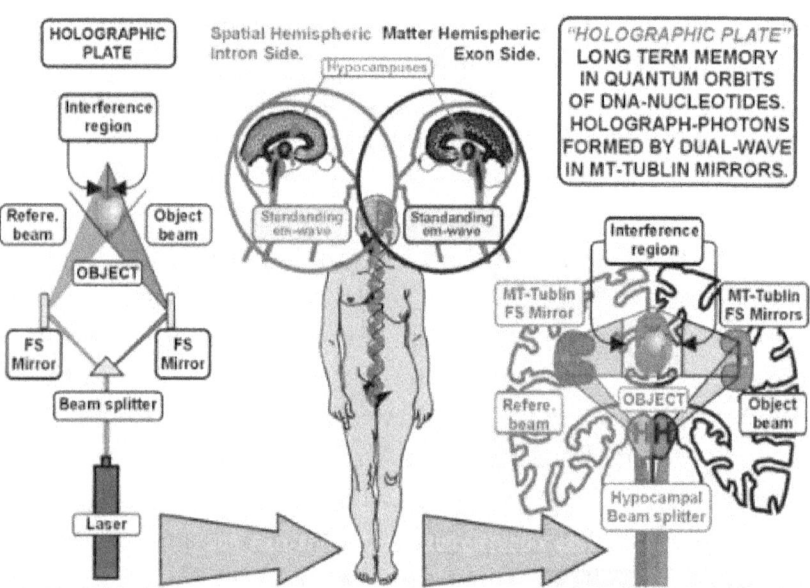

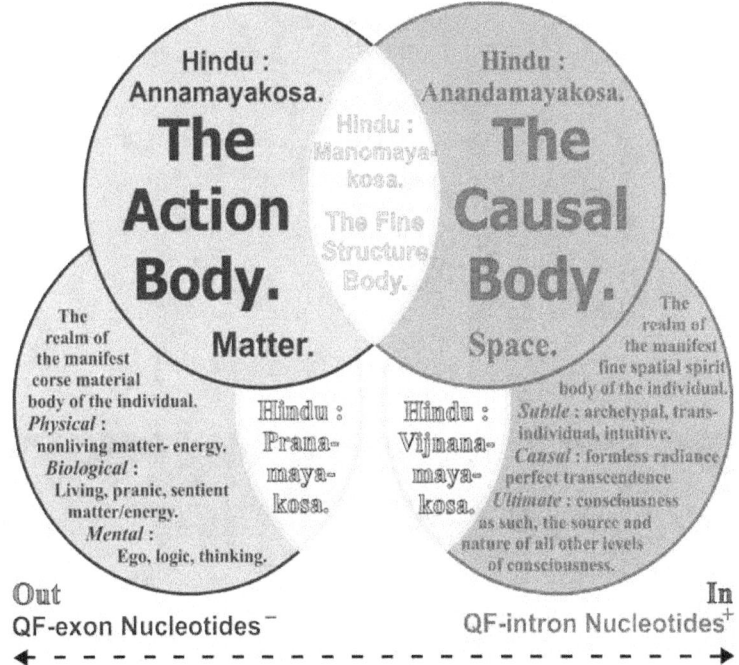

What creates the vortex of information?

Many cultures and religions use the swastika symbol in a variety of ways as you can see below. The word swastika comes from the Sanskrit svastika, which means "good fortune" or "well-being" and has been used for well over 5000 years. However the purpose of the swastika is not explained but I believe this is what creates the vortex of information as it rotates.

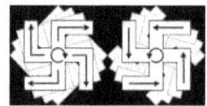

A right-facing swastika might be described as "clockwise" or "counter-clockwise".

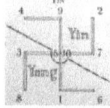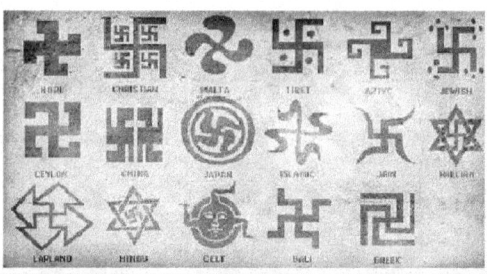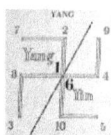

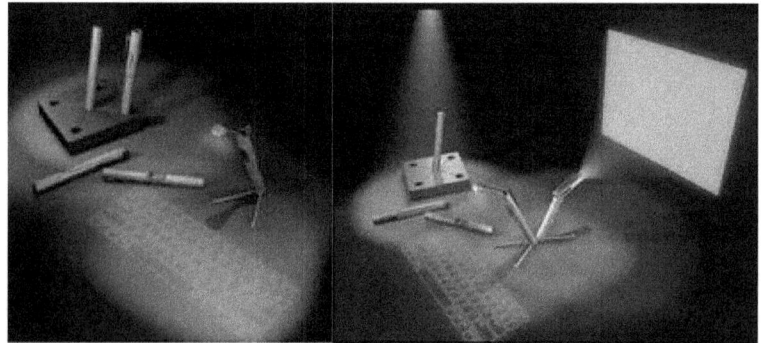

Everything is a hologram like the Blue tooth keyboard below. Light vibrates at frequency to create the illusion or hologram of the keyboard and screen. A grid matrix gives the perception of solidness and the high frequency flicker that you cannot see is the folding and recording of information and unfolding again to create the hologram as Walter Russell states in "The Secret of Light".

This is how the devices know that you have touched the key you require. You are the same as that keyboard, you are unfolding recording information and refolding, that is how you perceive time has moved on or receiving pain or any type of sensory perception.

Holographic Universe and Intelligent Design go hand in hand. Fractal and Euclidian geometry along with the Vedic measuring system are what allow the illusion to appear real. Hollywood movies always are mostly shot at 24 frames per second on film (24x 60 seconds = 1440 frames per minute x 60 minutes 86400fph) Our eyes only see a fraction of everything that is there so our universe mains a mystery to our mind and we only have our imagination left to conjure up the unseen. This information is from the documentary "the Colors of infinity" narrated by Arthur C Clark. If you are an observant person go out into the garden or park and find a tree or plant and pick out a section of the tree and try and find the exact same repeating pattern else

were on the tree and you will be able to see repeating patterns in trees and flowers. These two pictures are of the same tree just that the square is on a matching fractal section so you can observe the repeating pattern. The machine that generates this illusion is not any older than the day it was created. All reality and all dimensions are illusions; time only exists if you desire something to measure it.

As you read this, you may process the thought of who created this illusion of smoke and mirrors and for what purpose.

"If you want to see fear in a quantum physicist's eyes, just mention the words, 'the measurement problem.' The measurement problem is this: an atom only appears in a particular place if you measure it. In other words, an atom is spread out all over the place until a conscious observer decides to look at it. So the act of measurement or observation creates the entire universe." -Jim Al-Khalili, Nuclear Physicist

he concept of time and reality we have had imposed on us is farcical and the truth now needs to be represented into the light of day.

Taimni, in Science and Occultism, describes time and the universe as follows:
"The intermittent nature of time is a philosophical and scientific concept of the greatest significance and can be illustrated almost perfectly by the projection of a cinematographic picture on a screen. Although the projection produces the impression of a continuous series of events, we know that this continuity is merely an illusion produced by our inability to distinguish between the alternate periods of illumination and darkness. The manifested universe is similarly an intermittent phenomenon owing to the intermittent nature of time, which brings about changes in it. However, instants or moments of time succeed one another so rapidly that we cannot distinguish between the periods in which the universe appears and disappears alternately. The procession of events which we cognize through our sense-organs or instruments of cognition are thus seen as a continuous phenomenon but this continuity is illusory."

(Science and Occultism, pp.99-100)

The concept that light always travels at a constant speed is questionable to say the least as all matter is light/Aether vibrating at different frequencies to create the illusion of matter. In Quantum Mechanics have shown that a single electron can exist in two places at once. Common sense would tell you this would not be possible unless this single electron was a reflection of the first. Light does not travel vast distances it repeats itself from one light field to the next light field. All principles of nature are balanced and a lot modern science is void of the truth and creates false concepts to explain cause and effect. Where is the balance in that?

Walter Russell's mirrored cube is an example of how the endless universe is manifested.

Image from A NEW CONCEPT OF THE UNIVERSE by Walter Russell (1953.)

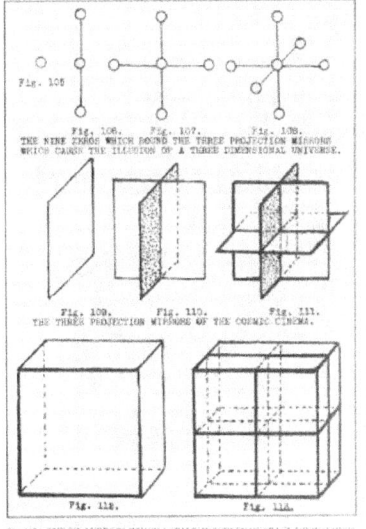
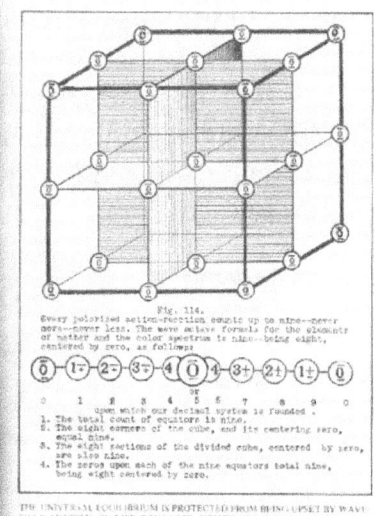

 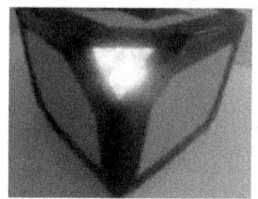 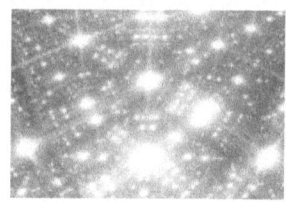

Walter Russell's Cubic Wave Field Model of Creation by Matt Presti http://www.youtube.com/watch?v=QSLp8SSIuII
Many have not gone far enough when explaining their concepts when stating we live in a holographic universe. Millions of people do not believe in God because of the false concepts and blatant lies told by various religions and alleged men of God. It should be known that balanced thinking in Krsna Consciousness is Gods thinking and not Concepts such as good and evil, which are based on duality. If you have maintained your life in Krsna consciousness, there should be no good or evil just the equilibrium of tranquility. Live your life and treat people in a manner you wish to be treated, "pay if forward". Do something out of kindness and love, give and do not expect a return.

The machine that creates a hologram requires a software language. The language of light.

Sir John Dee the original 007.

(1527-1608) was a genius, considered a philosopher and alchemist who captured the attention of the greatest minds throughout Europe.

Dee signed his letters with two circles symbolising his own two eyes and indicating that he was the secret eyes of the Queen. John Dee and his lucky number 7 could have come from many sources, the 7 core audible sound tones of creation KRYSTHL-A (Ka Ra Ya Sa Ta Ha La) or the 7 sided star or Osiris associated with Freemasonry.

The Vector/Chevron of N.A.S.A and the seven planets Saturn, Jupiter, Neptune, Uranus, Mars, Venus and Mercury. These are the seven of the Jewish candelabra.

Below is an example of how the language of light was discovered and is well worth reading and warranted further investigation. John Dee and Keven Bacon called it the language of the angles.

Images and excerpt from "The Complete Enochian Handbook" 1996 Vincent Bridges
Language of light Overview of system and components

The Ophanic Intelligences, the sentience of whirling Light, gave Dr Dee a powerful tool for leveraging reality. Imagine a magic tool box, small enough and portable enough to be scattered throughout the galaxy. The tool box contains "tools" designed to build a mechanism that functions as a combination radio set, life raft and emergency medical instrument. Included along with the tools is a DNA trigger coded instruction sheet.

Learning the alphabetic wave forms of symmetry set coherent sacred languages, such Hebrew, Greek, Sanskrit, etc., creates a stable psychic platform onto which the interlocking non-local dimensional structures of the Ophanic Language can be downloaded. This ensures that the tool box will be opened by someone with the spiritual perspective to use it. The information was originally given to John Dee because he could understand and respect the material. Mathers was able to expand upon the material for the same reasons.

Within the last five years, certain leaps forward in our understanding of the Ophanic rescue project have occurred. They point, with eerie synchronicity, toward the Enochian material of Dr Dee. Only now do we begin to see the outlines of the complete system and what it may be intended to accomplish.

The starting point is the alphabet itself. Arranged in the sixty-four characters used to spell the "title functions" of the letters, the alphabet creates a DNA sensitive matrix. Meditation on this pattern, and its evolution into a 16-cell 4D geometric figure, the dual of the Hyper-Cube, triggers a DNA coded process that allows us to emotionally and intellectually grok the Ophanic Language.

Once this is absorbed, the physical structure and psychic hardware of our "mechanism" can be constructed. The physical platform is the Holy Table. Its purpose is to support the floating focal point of the system, the Sigil of Truth and Orb Module.

The Table contains Ophanic letter patterns that generate an operational field domain that is continually unpacking higher sources into the local World of Action. It is insulated by small Sigils of Truth under each physical contact point to ensure that local negativity does not contaminate the system. Sort of like a surge protector on the power supply of a computer.

In fact the whole "mechanism" resembles a computer in its design and function. The Holy Table is an energy condenser charged by exposure to terrestrial currents. Think of it as the power supply and wiring patterns. Next came the psychic structure, the micro-chips, of the computing platform. These are the Elemental Tablets, formed into their hyper-cubic Temple structures, and the Tablet of Union that holds them in place and maintains continuity of process between diverse functions.

Hebrew unpacks only into a cube-octahedron, a truncation of a cubed octahedron. This "dymaxion" shape allows the surface of a sphere to map accurately onto 2 dimensions. The Ophanic language unpacks outward into higher dimensions.

The dodecahedron forms the transitional stage, a blend of Hebrew and Ophanic, while the higher unpacking of the hyper-cube forms the 24-cell 4D figure, which contains the 3D structure of both the dodecahedron and the icosahedron.

This structure is visualized as the environment in which the Holy Table, Sigil of Truth and the Orb operate. The Sigil of Truth is placed on the base of the Orb Module to act as a template or filter for truthful or coherent information exchange. The module itself supports this function by providing a direct link to a Tablet of Union balanced flow of energy. The Orb provides a link with the anchoring structures on Giza.

When the hardware is in place, then the operating system must be booted up, activating the processing routines. This is accomplished by vibrating the 18 Elemental Keys. These Keys also bring the Tablet of Union on line at the same time. Once these operations are in place, then the content rich extensions known as the Aires or Aethyrs can be activated.

The 30 Aethyrs form the constant localities symbolized by the 30 edges of both the dodecahedron and the icosahedron. These edges are contact zones where planetary and galactic forces inter-phase. The 49th Key of Silence acts as a modem connecting us with galactic information central.

Now our non-local computer is turned on and running. We have a choice of 93 programs that directly affect the earth and its immediate environment in space, and 700 programs for effecting individual and societal change. Our first act with the non-local system will be to call up Central and ask for more instructions before we begin to activate programs.

The goal of this handbook is to present the Ophanic material in a simple and direct way, with nothing left out or obscured by overly complex jargon. Only in this way can we actually expect to perform this complex spiritual operation. The clearer everyone is on what we are doing and how it works, even if they don't understand the details, the better the working will be.

B) Letters and Language
On March 26, 1582, early on in the transmission process, John Dee and Kelly were instructed to draw the characters they saw in an eight by eight square held up by the angel Ave. A faint yellow outline of an alphabetic character appeared on the page in front of Dee, which he then filled in with ink. The characters were somewhat similar to several magical scripts, including that described by Pantheus in his Voarchadumia, but more refined. They are also vaguely reminiscent of Ethiopian characters.

The 8 x 8 arrangement of the Enochian characters.

© 1996 Vincent Bridges and Darlene

The titles of the twenty-one letters emerged in three groups of seven; twenty characters spell seven titles in the first group, twenty-one in the second and twenty-three in the third, for a total of 64 characters. These titles are odd; they seem to have no relationship to the phonetic value of the letter.

The letters themselves have been demonstrated, by dowsing, to be tightly focused wave guides. The dowsing crystal forms a specific energy oscillation for each letter. The oscillation changes orientation with the letter, indicating that the wave guide of the letter is Omni-directional. Since each letter is asymmetrical, the complete group of twenty-one forms three variations on the seven-fold symmetry set. Only eighteen letters are used to spell out the titles of the letters; the three left out letters, B, Q, and Z, one from each of the three symmetry groups, form the three points on each face of an invisible tetrahedron. The missing or hidden 22nd letter, corresponding to the Greek Theta, is the invisible centre point of the tetrahedron.

Omni-directional 16 x 16 Grid.

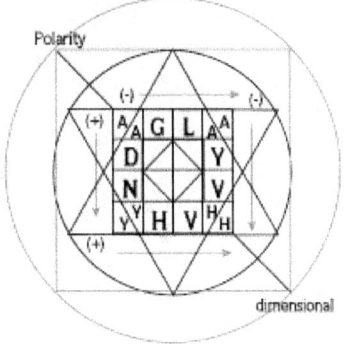

Four-letter God Name Octahedron

© 1996 Vincent Bridges and Darlene

Pyramid within a pyramid.

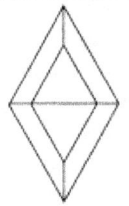

A Hyper-cube is the four-dimensional analogue of a three-dimensional cube.

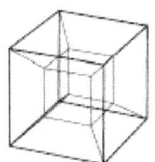

Spacial relationships of tetra- and octa- within cube.

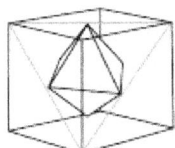

© 1996 Vincent Bridges and Darlene

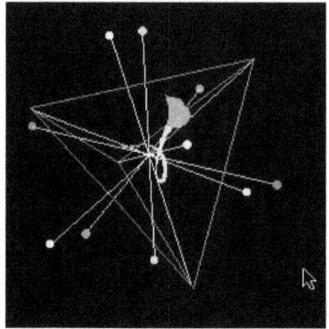

http://www.themeasuringsystemofthegods.com/Images/image675.gif

You can see an animation of this image at the link above.

Each 64 character square can be read from one of four directions, and the reading can start from each of the four corners. This produces 16 different symmetry sets or patterns. Each pattern, or way of reading the letter titles, produces 32 different sub-domains that are eight "names" or frequency signatures in each of the four directions. This gives us a grand total of 512, or 8 cubed, individual symmetry set domains.

The 16-cell 4D figure, a 4D octahedron, or a pyramid within a pyramid, has the same relationship to the hyper-cube as a 3D octahedron has to a cube. Just as the vertices of the octahedron form the faces of the cube, so do the vertices of the 16-cell hyper-octahedron form the cells of the hyper-cube. This arrangement is familiar from the development of the hyper-cube within the Meta-Programming Protocol.

In that instance, the 16 Hebrew letters used to spell the four letter names of God in the Lesser Ritual of the Pentagram are used to form a 4D octahedron; eight letters, A, D, N, Y, G, V, L, and H, form the outer surface of the 3D octahedron. The repeated letters, A, (3) Y, (1) V, (2) and H(2) form the inner cells of the 4D octahedron.

The [...] vertices
of this [...] 4D

**16-Cell 4D FIGURE &
LANGUAGE
GRID EQUIVALENT**

8 vertices	=	ranking of sub-domains
24 Faces	=	16 symmetry sets plus 8 categories of sub-domain
32 Edges	=	Number of sub-domains per symmetry set
16 Cells	=	Number of symmetry sets

octahedron form the cells of the hyper-cube created by the four pentagrams and the four god names. The archangelic names form the four faces and hidden centre point of a tetrahedral symmetry generator that animates the hyper-cube.

Each of the basic sixteen symmetry patterns of the 64 square Ophanic alphabetic structure forms a cell of the 4D octahedron. The eight patterns formed on the horizontal or the vertical are the eight surface faces that show in a 3D octahedron. The eight diagonal patterns form the inside cells of the 4D octahedron.

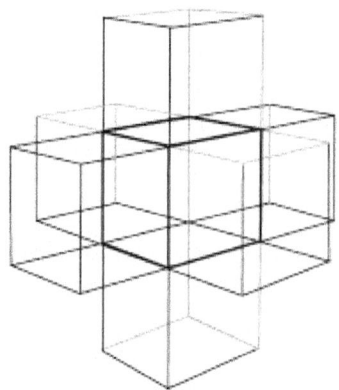

Three-dimensional cross formed by placing six cubes on the faces of the seventh.

© 1996 Vincent Bridges and Darlene

Within each cell, the 32 sub-domains form the faces of the 4D figure; the 16 directions and the eight categories of sub-domain form the 24 edges of the 16-cell figure. The eight vertices are formed by the same ranking as the categories of sub-domain, which is either eight across, eight down, eight names with one to eight letters, etc.

So far, we have developed a dimensional structure similar to that of Hebrew; a spinning tetrahedron throwing off symmetry angles inside a 4D octahedron that turns itself inside out to form the hyper-cube of the pentagram ritual. The Ophanic alphabet of symmetry also uses a tetrahedron (focused on the vertices rather than the faces) and generates a 4D octahedron from its symmetry sets and sub-domains.

However, instead of turning inside out to form a hyper-cube, the Ophanic language crystal 4D octahedron folds outward to form six hyper-cubes, one unfolding from each face of the Hebrew and Ophanic Cube of Space. (Both Hebrew and the Ophanic Letters form the faces, edges, axis of orientation and central point of a space defining cube.) Each of these outwardly unfolding hyper-cubes is formed from one of the four Elemental Tablets.
Unlike the Hebrew 16-cell 4D figure, which uses only eight of 22 letters and delineates no further internal 4D components, but is used at every working, the Ophanic 4D octahedron is fully developed structurally and never used directly in the Ophanic Language or system.
It seems to be a mental construct designed to be absorbed by the unconscious and then used to leverage the interactions of the system at the level of the DNA.

It is helpful to think of the Ophanic 4D octahedron as a seed containing all the possible DNA derived probability waves. In other words, the literal matrix of creation.

The Ophanic language developed from the squares described in the Liber Logaeth, or "Book of the Speech of God." Some of this early form of the Ophanic language resembles verb tense and other grammatical drills. None of this was translated directly into English, although Dee ventured a few interpretations. At any rate, a large number of squares were generated in this trance-like way.

In 1584, John Dee and Kelly received instructions on how to convert the squares of some of the tablets in the Liber Logaeth into invocations or calls. This time, the angels provided a translation. The language that emerged had English phonology and grammar, but the roots and other vocabulary elements are not directly derivable from English, Latin, Greek, or Hebrew. Some roots are similar to Sanskrit and Egyptian, but some are simply alien in construction and pronunciation.

The Keys or Calls contain the purest example of the Ophanic Language that we have. The names formed from the Tablets are more random in appearance, and are much more difficult to translate. The Keys seem to provide a literary structure, the grammar and syntax, which are needed to operate the more complex frequency signatures embedded within the structure.

Summary of system and structure

The Ophanim delivered to Dr John Dee and Edward Kelly a complex system of magical science that is both inclusive and inter-dimensional. It seems to function as a computer system, with hardware, software and programs to run, but one composed of geometrical shapes and frequency matrices. It is impossible to go very far in understanding the Ophanic material without coming to the conclusion that this is proof of contact with a higher level of intelligence.

Basic symmetry set alignment and dimensional projection of geometric forms are accomplished by the Hebrew language operations. The Hypercube, and its dual the 16-cell, become the basic building block of the entire structure.

I would like to thank Vincent Bridges for deciphering this information and presenting it into a format for everyone to appreciate. Please read his entire book to get more information on this interesting subject.

A simple explanation of hyper-cubes and other dimensions using geometry. The following images are from that great book "A Primer of Higher Space (The Fourth Dimension) By Claude Bragdon 1913.

THE GENERATION OF CORRESPONDING FIGURES IN ONE-, TWO-, THREE-, AND FOUR-SPACE.

FIG. 1. ———

THE LINE: A 1-SPACE FIGURE GENERATED BY THE MOVEMENT OF A POINT, CONTAINING AN INFINITE NUMBER OF POINTS, AND 2 FORM ITS BOUNDARIES

FIG. 2.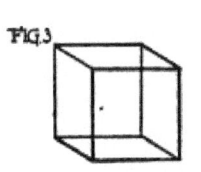

THE SQUARE: A 2-SPACE FIGURE GENERATED BY THE MOVEMENT OF A LINE IN A DIRECTION PERPENDICULAR TO ITSELF TO A DISTANCE EQUAL TO ITS OWN LENGTH IT CONTAINS AN INFINITE NUMBER OF LINES, AND IS BOUNDED BY 4 LINES AND 4 POINTS.

FIG. 3.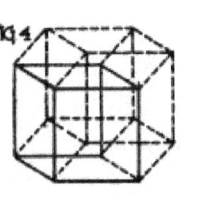

THE CUBE: A 3-SPACE FIGURE OR 'SOLID,' GENERATED BY THE MOVEMENT OF A SQUARE, IN A DIRECTION PERPENDICULAR TO ITS OWN PLANE, TO A DISTANCE EQUAL TO THE LENGTH OF THE SQUARE THE CUBE CONTAINS AN INFINITE NUMBER OF PLANES (SQUARES) AND IS BOUNDED BY 6 SURFACES, 12 LINES AND 8 POINTS

FIG. 4.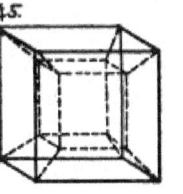

FIG. 5.

THE TESSERACT, OR 'TETRA-HYPERCUBE': A 4-SPACE FIGURE GENERATED BY THE MOVEMENT OF A CUBE IN THE DIRECTION (TO US UNIMAGINABLE) OF THE 4TH DIMENSION. THIS MOVEMENT IS EXTENDED TO A DISTANCE EQUAL TO ONE EDGE OF THE CUBE AND ITS DIRECTION IS PERPENDICULAR TO ALL OUR 3 DIMENSIONS AS EACH OF THESE 3 IS PERPENDICULAR TO THE OTHERS. THE TESSERACT CONTAINS AN INFINITE NUMBER OF FINITE 3-SPACES (CUBES) AND IS BOUNDED BY 8 CUBES, 24 SQUARES, 32 LINES AND 16 POINTS.

NOTE: FIGURE 4 IS A SYMBOLIC REPRESENTATION ONLY — A SORT OF DIAGRAM — SUGGESTING SOME RELATIONS WE CAN PREDICATE OF THE TESSERACT. FIGURE 5 IS A REPRESENTATION DRAWN ON A DIFFERENT PRINCIPLE IN ORDER TO BRING OUT A DIFFERENT SET OF RELATIONS.

CORRELATIONS, IN FORM AND SPACE, OF SOME PROPERTIES OF ABSTRACT NUMBER

A NUMBER MULTIPLIED BY ITSELF GIVES THE SECOND POWER OF THAT NUMBER, COMMONLY CALLED ITS "SQUARE" BY REASON OF ITS CLOSE RELATION TO THE GEOMETRICAL SQUARE WHOSE SIDE CONTAINS THE GIVEN NUMBER OF UNITS OF LENGTH.

THE 3 LINEAR UNITS

FIG. 1

SIDE OF SQUARE 3 LINEAR UNITS, AREA OF SQUARE 9 SQUARE UNITS

THE SECOND POWER OF 3 IS 3 TIMES 3, OR 9. FIG. 1 REPRESENTS A GEOMETRICAL SQUARE WHOSE SIDE IS 3 UNITS IN LENGTH, SAY 3 INCHES. THE AREA OF THE SQUARE WILL OBVIOUSLY BE 9 SQUARE INCHES.

FIG. 2

EDGE OF CUBE 3 LINEAR UNITS, FACE OF CUBE 9 SQUARE UNITS, VOLUME OF CUBE 27 CUBIC UNITS

LET US NOW BUILD UP FROM THE SQUARE TO A HEIGHT OF 3 INCHES THE CUBE REPRESENTED IN FIG. 2. THE SOLID WILL OBVIOUSLY CONTAIN 3 X 3 X 3 = 27 CUBIC INCHES.

BY ANALOGY WITH THE GEOMETRICAL FIGURE, THE NUMBER 27, THE 3RD POWER OF 3, IS CALLED IN ARITHMETIC THE "CUBE" OF 3.

NOW THE 4TH, 5TH AND HIGHER POWERS OF A NUMBER ARE COMMONPLACES OF ARITHMETIC. WHAT DO SUCH HIGHER POWERS MEAN IN GEOMETRY?

WE CANNOT MAKE COMPLETE PHYSICAL REPRESENTATION OF 4-DIMENSIONAL SOLIDS IN OUR 3-SPACE, JUST AS WE CANNOT CONSTRUCT A CUBE IN A PLANE SURFACE, BUT WE CAN MAKE DIAGRAMS OF HYPERSOLIDS, AND THE PROPERTIES OF MANY SUCH FIGURES IN HYPERSPACE ARE WELL KNOWN, HAVING BEEN DEMONSTRATED LIKE PROPOSITIONS IN EUCLID.

HYPERSPACE IS THUS MATHEMATICALLY REAL, AND THE MASTER MINDS OF SCIENCE CONSIDER IT TO BE PHYSICALLY POSSIBLE (LORD KELVIN AND OTHERS).

A 2-SPACE UNIT   A 3-SPACE UNIT   A 4-SPACE UNIT

THE DEVELOPMENT OF A UNIT OF 2, 3, AND 4 SPACE INTO THE NEXT LOWER SPACE AND THEIR EXPRESSION IN AND BY MEANS OF UNITS OF THOSE LOWER SPACES

IF THE BOUNDING LINES OF THE SQUARE A-B-C-D WERE MADE OF A CONTINUOUS WIRE, AND IF THAT WIRE WERE CUT AT D, THE BOUNDARY COULD THEN BE BENT DOWN INTO LINE WITH A-B FORMING A ONE-DIMENSIONAL FIGURE OF FOUR LINEAR UNITS—THE ORIGINAL LINEAL UNIT A-B HAVING ONE LINEAL UNIT AT EACH END OF IT AND AN EXTRA ONE BEYOND AT ONE END

IF THE CUBE A-B-C-D—G WERE MADE OF A CONTINUOUS SHEET OF TIN AND IF THAT SHEET WERE CUT ALONG CERTAIN LINES FORMED BY INTERSECTING FACES, THE WHOLE COULD BE FOLDED DOWN TO FORM A TWO-DIMENSIONAL FIGURE OF SIX SQUARES—THE SQUARE A-B-C-D HAVING A SQUARE ON EACH SIDE OF IT AND ONE BEYOND ON ONE SIDE

SIMILARLY, IF THE TESSERACT (REPRESENTED BY THE DIAGRAM) WERE MADE OF SOLID WOOD AS TO ITS BOUNDING CUBES AND IF THIS WOOD WERE CUT THROUGH THE APPROPRIATE PLANES, THE CUBES COULD, BY ANALOGY, BE FOLDED DOWN TO FORM A THREE DIMENSIONAL FIGURE OF EIGHT CUBES

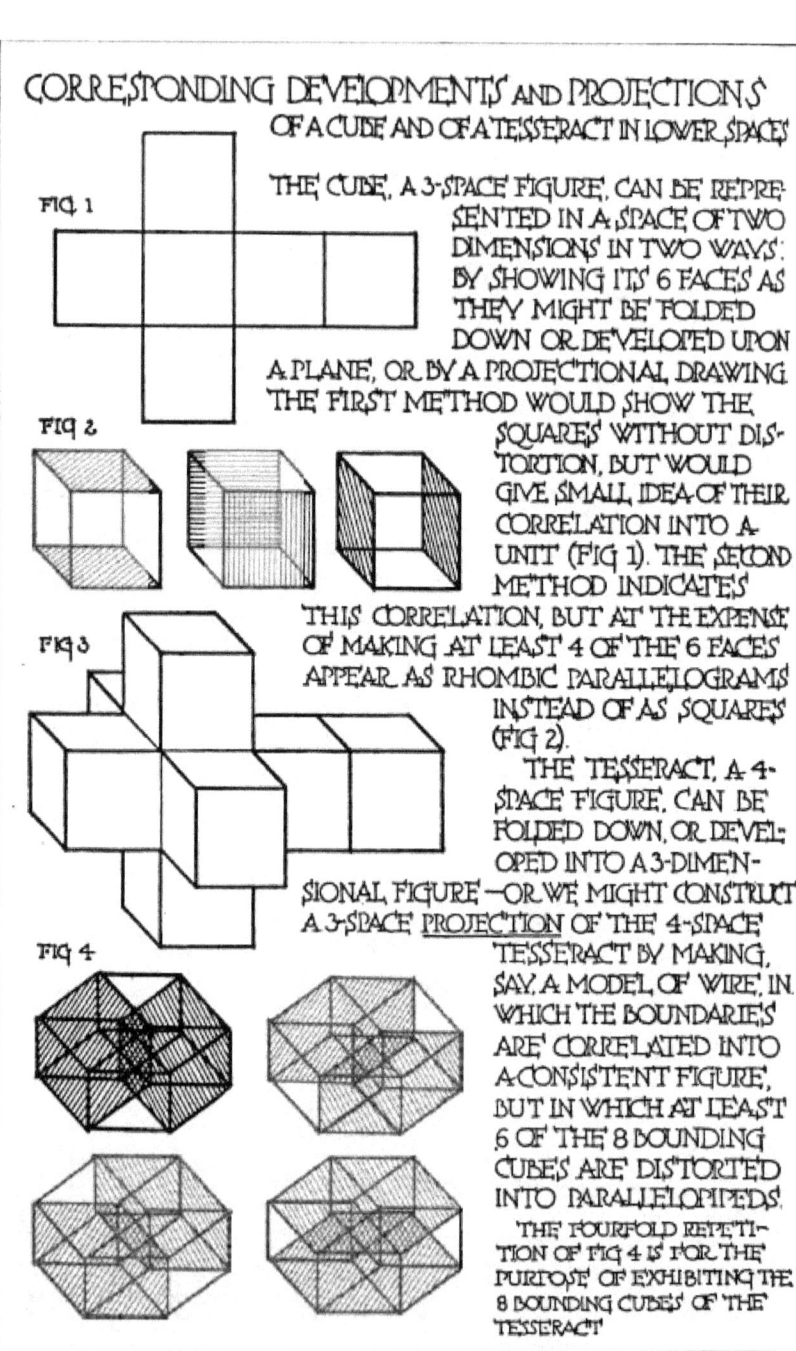

CORRESPONDING DEVELOPMENTS AND PROJECTIONS OF A CUBE AND OF A TESSERACT IN LOWER SPACES

THE CUBE, A 3-SPACE FIGURE, CAN BE REPRESENTED IN A SPACE OF TWO DIMENSIONS IN TWO WAYS: BY SHOWING ITS 6 FACES AS THEY MIGHT BE FOLDED DOWN OR DEVELOPED UPON A PLANE, OR BY A PROJECTIONAL DRAWING. THE FIRST METHOD WOULD SHOW THE SQUARES WITHOUT DISTORTION, BUT WOULD GIVE SMALL IDEA OF THEIR CORRELATION INTO A UNIT (FIG 1). THE SECOND METHOD INDICATES THIS CORRELATION, BUT AT THE EXPENSE OF MAKING AT LEAST 4 OF THE 6 FACES APPEAR AS RHOMBIC PARALLELOGRAMS INSTEAD OF AS SQUARES (FIG 2).

THE TESSERACT, A 4-SPACE FIGURE, CAN BE FOLDED DOWN, OR DEVELOPED INTO A 3-DIMENSIONAL FIGURE—OR WE MIGHT CONSTRUCT A 3-SPACE <u>PROJECTION</u> OF THE 4-SPACE TESSERACT BY MAKING, SAY, A MODEL OF WIRE IN WHICH THE BOUNDARIES ARE CORRELATED INTO A CONSISTENT FIGURE, BUT IN WHICH AT LEAST 6 OF THE 8 BOUNDING CUBES ARE DISTORTED INTO PARALLELOPIPEDS.

THE FOURFOLD REPETITION OF FIG 4 IS FOR THE PURPOSE OF EXHIBITING THE 8 BOUNDING CUBES OF THE TESSERACT.

THE REPRESENTATION AND ANALYSIS OF THE TESSERACT, OR FOUR-DIMENSIONAL CUBE BY A METHOD ANALOGOUS TO THAT EMPLOYED IN MAKING A PARALLEL PERSPECTIVE

FIG 1.

A GLASS CUBE, HELD DIRECTLY IN FRONT OF THE EYE, WILL APPEAR AS SHOWN IN THE ACCOMPANYING DRAWING. THIS BEING A PLANE FIGURE OF TWO DIMENSIONS-MIGHT HAVE BEEN PRODUCED BY DRAWING ONE SQUARE INSIDE OF ANOTHER AND THEN CONNECTING THE CORRESPONDING CORNERS. THIS COULD BE DONE WITHOUT ANY THOUGHT OF THREE DIMENSIONS, YET ON THIS PLANE FIGURE MANY OF THE PROPERTIES OF THE CUBE CAN BE STUDIED. BY COUNTING THE FOUR-SIDED FIGURES, WHICH WE FIND TO BE SIX, WE LEARN THE NUMBER OF FACES OF THE CUBE. BY COUNTING THE NUMBER OF CORNER POINTS, WHICH ARE EIGHT, WE LEARN THE NUMBER OF THE CORNERS OF THE CUBE. BY COUNTING THE LINES, WHICH ARE TWELVE, WE LEARN THE NUMBER OF EDGES OF THE CUBE.

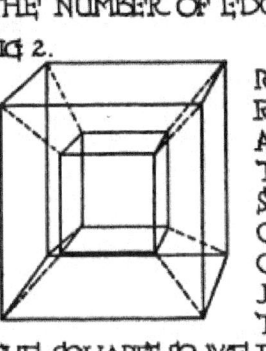

FIG 2.

IN THE SAME WAY THAT FIGURE 1 REPRESENTS THE CUBE, FIGURE 2 REPRESENTS THE THREE-DIMENSIONAL FORM CORRESPONDING TO THE TESSERACT. JUST AS WE DREW A SMALLER SQUARE INSIDE OF A LARGER ONE, SO WE REPRESENT A SMALLER CUBE INSIDE OF A LARGER CUBE. AND JUST AS WE DREW LINES JOINING THE CORRESPONDING CORNERS OF THE SQUARES, SO WE FORM PLANES JOINING THE CORRESPONDING EDGES OF THE CUBES. TO FIND THE NUMBER OF CUBIC BOUNDARIES OF THE TESSERACT, WE COUNT THE LARGE OUTER CUBE, THE SMALL INNER CUBE, AND THE SIX SURROUNDING SOLIDS—EACH A DISTORTED CUBE—EIGHT IN ALL. A FURTHER STUDY OF THE FIGURE DISCOVERS 24 PLANE SQUARE FACES, 32 EDGES, 16 CORNER POINTS.

EACH "SPACE," OR DIMENSIONAL ORDER, CONTAINS AN INFINITY OF SPACES OF DIMENSIONS FEWER BY ONE, AND IS ITSELF ONE OF AN INFINITE NUMBER OF SIMILAR SPACES CONTAINED WITHIN A SPACE OF DIMENSIONS GREATER BY ONE

FIG 1.

A LINE CONTAINS AN INFINITE NUMBER OF POINTS (FIG 1).

FIG 2.

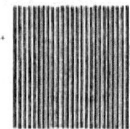

A PLANE CONTAINS AN INFINITE NUMBER OF LINES—1-SPACES (FIG 2).

A SOLID CONTAINS AN INFINITE NUMBER OF PLANES—2-SPACES (FIG 3).

BY ANALOGY, AN INFINITE NUMBER OF SOLIDS—3-SPACES—WOULD BE CONTAINED WITHIN A HYPER-SOLID—A 4-SPACE.

FIG 3.

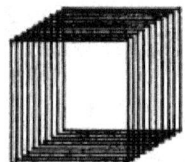

IF, AS PHILOSOPHERS AFFIRM, THE VISIBLE WORLD EXISTS ONLY IN, AND FOR CONSCIOUSNESS—IF IT IS BUT THE "PERCEPTION OF A PERCEIVER"—THEN FOR EACH CONSCIOUS PERSON THERE EXISTS A DIFFERENT WORLD.

IT FOLLOWS LOGICALLY THAT THESE COUNTLESS PERSONAL CONSCIOUSNESSES IN WHICH THE THREE-DIMENSIONAL PERCEPTION OF THE WORLD INHERES, MAY BE THOUGHT OF AS SO MANY 3-SPACES GOING TO FORM A HIGHER, OR FOUR-DIMENSIONAL UNITY—THE CONSCIOUSNESS OF HUMANITY AS A WHOLE. FOR IT IS CLEAR THAT HUMANITY IS HIGHER-DIMENSIONAL IN RELATION TO THE INDIVIDUAL MAN IF WE CONSIDER HUMANITY IN ITS TOTALITY, IT HAS POWERS OF WHICH NO SINGLE HUMAN BEING IS POSSESSED. IT IS BOTH OLD AND YOUNG, YET DEATHLESS; IT IS IN ALL PLACES AT ONCE, IT SEES ALL OBJECTS, HEARS ALL SOUNDS, THINKS ALL THOUGHTS, EXPERIENCES ALL SUFFERINGS, ALL DELIGHTS. NOW SUPPOSE A MAN TO DWELL CONSTANTLY IN THE THOUGHT OF THIS HUMANITY, TO IDENTIFY ALL HIS INTERESTS WITH ITS LARGER INTERESTS, IS IT NOT THINKABLE THAT HE MIGHT TRANSCEND THE PERSONAL LIMITATION, AND MERGE HIMSELF INTO THE LARGER CONSCIOUSNESS OF WHICH HE HAS ALL THE WHILE BEEN A PART?

THE BOUNDARY BETWEEN TWO ADJACENT PORTIONS OF ANY SPACE IS, IN GENERAL, A SPACE OF DIMENSIONS FEWER BY ONE —

FOR THE DISCUSSION OF PHYSICAL REALITIES WE ARE TO CONCEIVE EACH KIND OF SPACE AS POSSESSING AN INFINITESSIMALLY SMALL EXTENSION IN THE NEXT HIGHER DIMENSION

FIG 1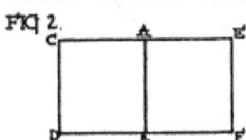

A POINT, OR 0-SPACE DIVIDES A LINE, OR 1-SPACE INTO TWO PARTS (FIG 1) FOR OUR PHYSICAL REASONINGS WE MAY TAKE FOR THE POINT A CIRCLE OF 1/200,000,000 OF AN INCH DIAMETER — ABOUT THAT OF A MOLECULE.

THE LINE A—B (FIG 2), A 1-SPACE, FORMS THE BOUNDARY BETWEEN TWO ADJACENT PARTS OF THE PLANE OR 2-SPACE C D F E. NOTE THAT AS THE DIVIDED LINE (FIG 1) IS HIGHER SPACE TO THE DIVIDING POINT, SO THE PLANE IS HIGHER SPACE TO THE LINE THAT DIVIDES IT AND BOUNDS THE TWO ADJACENT PORTIONS

AGAIN: THE PLANE G (FIG 3) A 2-SPACE SEPARATES AND MUTUALLY BOUNDS TWO ADJACENT PORTIONS OF THE SOLID 3-SPACE H I K J N L M O

THEREFORE, WE HAVE AT LEAST RIGID ANALOGY TO JUSTIFY US IN SAYING THAT OUR 3-SPACE DIVIDES HYPERSPACE OF 4 DIMENSIONS, AND MUTUALLY BOUNDS TWO ADJACENT PORTIONS THEREOF.

THE POINT P IS WHOLLY CONTAINED IN THE LINE, EVERY POINT OF THE LINE A B IS IN CONTACT WITH ITS HIGHER SPACE, THE PLANE; EVERY PART OF THE PLANE G IS IMMERSED IN ITS HIGHER SPACE, THE SOLID TOUCHING EVERY POINT OF THE PLANE. SIMILARLY OUR 3-SPACE, AS A BOUNDARY IMMERSED IN 4-SPACE, MUST UNDENIABLY BE IN CONTACT AT EVERY POINT WITH THAT 4-SPACE THAT IS, THE INNERMOST PARTS OF OUR SOLIDS ARE AS OPEN TO TOUCH FROM 4-SPACE AS ARE ITS BOUNDARIES TO US.

These last seven images where from "A Primer of Higher Space" (The Fourth Dimension) By Claude Bragdon 1913. If you are interested in knowing more this is an excellent book for the layman to understand.
"Saturn" and origin of "SATURN CUBE" Worship

The Kaaba, the cube building in Mecca and the Christians cross which itself is a folded out cube. I will let the reader draw their own conclusion here to why they chose these symbols.

"Israel" Many esoteric researchers affirm that the name Is-Ra-El is the combination of the names of ancient pagan deities Isis, Ra and El.

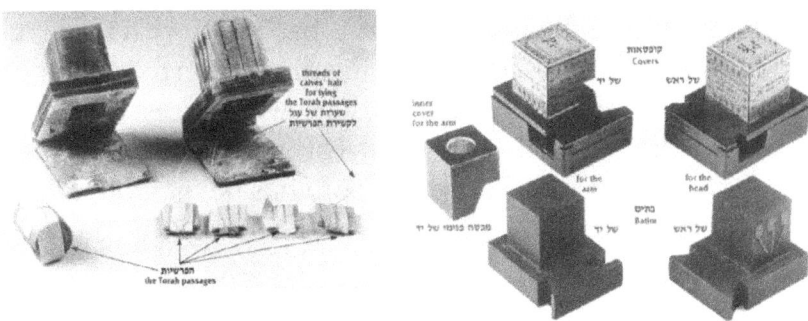

http://ott.co.il/tefillin/the-batim-boxes/

The Tefillin is also called phylacteries from Ancient Greek phylacterion, meaning "to guard, protect") are a set of small black leather boxes containing scrolls of parchment inscribed with verses from the Torah, which are worn on the head by Jews during weekday morning prayers.

http://en.wikipedia.org/wiki/Tefillin

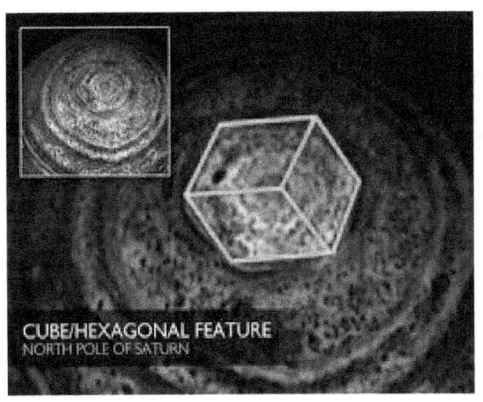

North Pole of Saturn hexagon cloud pattern.

This Cube within the hexagon cloud has to do with the ancient worship of El. Semitic civilizations referred to the god Saturn as "El". The supreme deity was represented by a black cube. We can find examples of the cube across the world. The practise of Saturn worship has never stopped and its rites are still imbedded in ceremonies to this day.

Transmutation.

At the Westinghouse laboratories Walter Russell used a device to transmute water into 17 different elements.

Quote from Walter Russell.

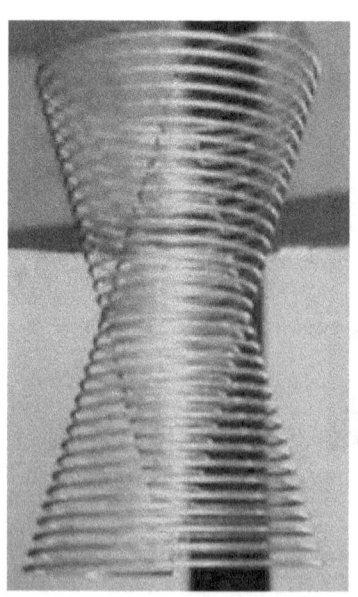

"The entire substance of universal mind is thinking in varying but orderly rhythmic meter. (Vedic metric system)The meter of the universal thinking is measurable in its orderliness throughout the entirety of the universal substance. The tempo of the cosmic, rhythmic meter of thinking is absolute." (VEDIC METRIC)

One reason many people can't recreate the experiments of Tesla, Keely, Russell and many others is that the wrong science and thinking is being applied. You must incorporate the correct geometry using harmonic proportion and use a measuring system that has been incorporated into the software of God's creation. If you read the above quote transmutation that occurs in nature is adjusting the rhythmic tempo using a rhythmic meter. To put it simply, use musical tones. Once again we find our way back to the cube. The illusion of life and all substances comes from light played as music generated with a mirrored cube.

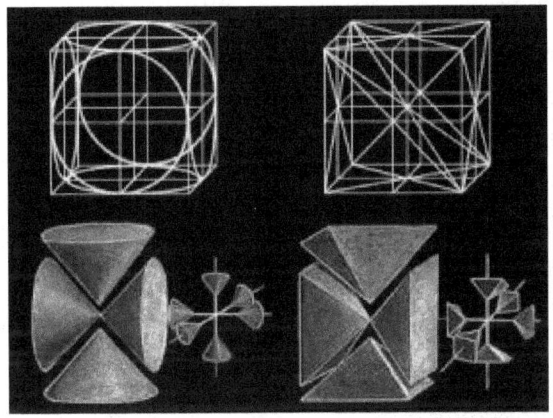

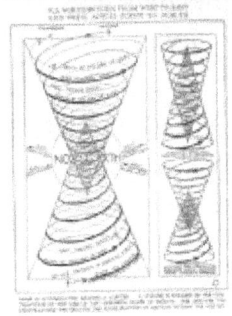

So to recreate this process of transmutation it would make sense to design a device that mimics nature. Using this simple example to demonstrate Russell's beliefs I am sure many of you can in vision natures process of transmutation.

Why do you think they have kept transmutation a secret from everyone?

Fractals

If you examine that mathematical equation of the Mandelbrot set and how Fractal Geometry is part of the building blocks for everything you should see the repeating patterns within nature.

The Mandelbrot set: $z = z^2+c$ where z is a complex variable and c is a complex constant (as known a "complex number" The major shape-varieties of the Julia Sets correspond to regions in the Mandelbrot Set. This formula with a few variants can create images of ferns as pictured to the left to everything found in nature.

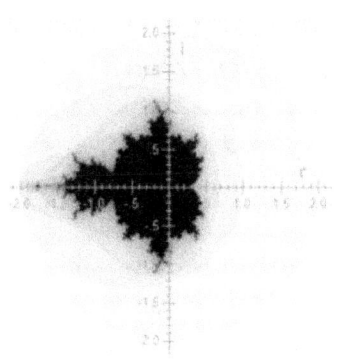

In the Mandelbrot set, nature (or is it mathematics) provides us with a powerful visual counterpart of the musical idea of 'theme and variation': the shapes are repeated everywhere, yet each repetition is somewhat different. It would have been impossible to discover this property of iteration if we had been reduced to hand calculation, and I think that no one would have been sufficiently bright or ingenious to 'invent' this rich and complicated theme and variations. It leaves us no way to become bored, because new things appear all the time, and no way to become lost, because familiar things come back time and time again. Because this constant novelty, this set is not truly fractal by most definitions; we may call it a borderline fractal, a limit fractal that contains many fractals. Compared to actual fractals, its structures are more numerous, its harmonies are richer, and its unexpectedness is more unexpected - **Benoit Mandelbrot**.

The growth of all plant is far from random. They follow given patterns. The branch growth from a tree depends on its species and individual genetic disposition. And as a tree grows its general pattern will stay the same but the length of branches will change depending on the development of the tree crown. The age of the tree and the different light conditions it's exposed to influence the tree's growth. These images of an Elm tree and European beech branches where created using a fractal generator

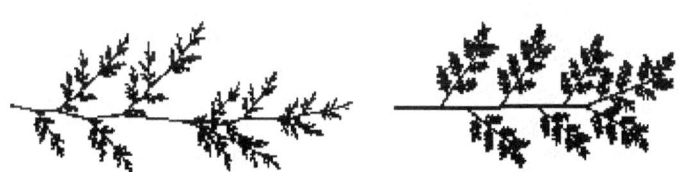

Quote from "A NEW CONCEPT OF THE UNIVERSE by Walter Russell (1953.)" When talking about misconceptions of modern science.

"Failure to recognise that this universal body of moving matter has been created by some power outside of itself has led science to conclude that the energy which created matter is within itself. Even more erroneous is the conclusion that energy is a condition of matter such as heat."

What Walter Russell meant by "some power outside of itself" is this holographic universe of mirrors is generated by an external artificial source. Holograms are projections of light from a device which is manufactured for that express purpose. What is thought of as God is the software for an ever creating illusion of the ONE mind.

What is being explained here is that we are living in a matrix. If you choose to believe, or not, is your choice as you have free will as God intended. In saying this, is your thinking balanced if you ignore Gods creation?

Sound will heal you

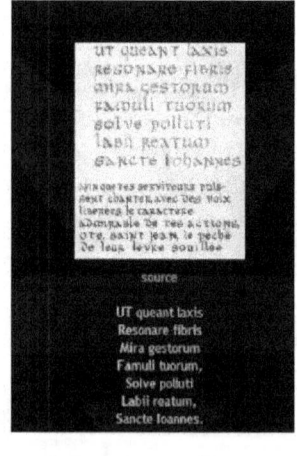

UT — 396 Hz — Liberating Guilt and Fear
RE — 417 Hz — Undoing Situations and Facilitating change
MI — 528 Hz — Transformation and Miracles (DNA repair)
FA — 639 Hz — connecting/Relationships
SOL — 741 Hz — Awakening Intuition
LA — 852 Hz — Returning to Spiritual Order

Frequency Machines where used to cure many diseases in the 18th and early 19th century. The Solfeggio healing frequencies were once in Webster's Dictionary but has been suppressed. Modern science is just starting to use them now and this original scale can awaken and heal you.

For example, the third note, frequency 528, relates to the note MI on the scale and derives from the phrase "MI-ra gestorum" in Latin meaning "miracle." Stunningly, this is the exact frequency used by genetic biochemists to repair broken DNA – the genetic blueprint upon which life is based! These frequencies can be listened to through Tone Generator using a square wave but a tuning would be better. There are more frequencies, nine in total but I am unaware of what they are for.

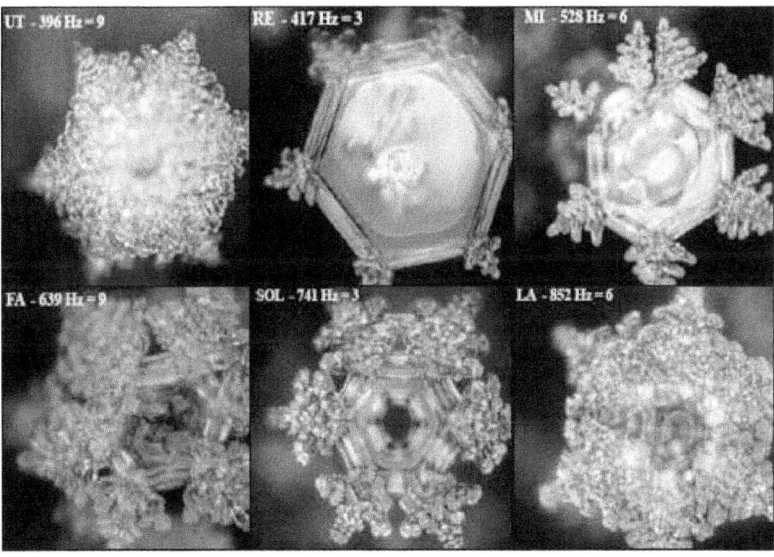

Each musical tone creates different geometrical patterns in the water crystals Photos From: powerofharmony.tripod.com

Vedic Physics and the Fifth Element.

Earth, air, fire, water and Aether

The fifth point on the Star of David

Let us conceive now, in accordance with the teachings of both Modern Science and Occult Science, that all Space is filled with Something, the most philosophical term for which would be Primordial Substance, because the term Substance means literally that which sub-stands, or stands under, as a root or basic Principle. It is that from which the phenomenal Universe is evolved, that by reason of which it exists or comes into existence (from ex, out, and sisto, to stand), i.e., that which appears or is manifested of an otherwise concealed Cause or Principle. Considering this Principle, however, in a somewhat more restricted sense, as merely the Root of Matter, we may give other names to it; and the name which Modern Science gives to it is Ether.

The Ancients also called it Aether; only their Aether was not the Ether of Modern Science, but something very much further removed from physical Plane phenomena. We may retain the distinctive spelling of the word in order to indicate the difference.

Now it is not difficult to postulate the existence of "one universal substance, perfectly homogeneous and continuous and simple in structure, extending to the furthest limits of space of which we have any knowledge, existing equally everywhere" (Sir Oliver Lodge).

In The Secret Doctrine there are three distinctive names or terms used to connote the Root Principle of the Phenomenal Universe, or what we have hitherto termed Primordial Substance. These three terms are respectively, Mulaprakriti, Akasha, and Aether.

They are all practically the same in their physical application, though there are shades of metaphysical difference which are not of much importance to us in this work. Mulaprakriti is a Sanscrit term meaning literally the Root of Nature (Prakriti), or Matter. Akasha is also a Sanscrit word. It is defined in the Theosophical Glossary as "The Universal Space in which lies inherent the eternal Ideation of the Universe in its ever-changing aspects on the planes of matter and objectivity, and from which radiates the First Logos, or expressed thought." It is also defined as, "The subtle, supersensuous spiritual essence which pervades all space; the primordial substance erroneously identified with Ether." Aether is a Greek word, and appears to be more specially associated with the idea of Light — Primordial Light, the Light of the Logos, not our physical light — than with anything else.

"THE PHYSICS OF THE SECRET DOCTRINE" BY WILLIAM KINGSLAND. 1910.

"Light is the living substance of Mind in action. It is the creating principle of the ONE substance. The ONE substance is the etheric "spiritual" substance of the ONE universal Mind. The entire "created" universe of all that is, ever has been, or ever will be, is but the ONE substance in motion, light." [Russell, The Universal One]

The Primal ONE, the Pure Light, without beginning and without end.

The word "aether" means "shine" in Greek, and the fundamental reality of such an unseen, fluid like source of universal energy has long been the knowledge possessed by of the world's secret mystery schools. The works of Greek philosophers Pythagoras and Plato discussed it at great length, as did the Vedic scriptures of ancient India, referring to it as "prana" and "Akasha."

Definition from
http://www.newworldencyclopedia.org/entry/Maya

Maya (Sanskrit māyā, from mā "not" and yā "this") is a term denoting three interrelated concepts: 1) power which enables those in its possession, most often gods, to produce forms in the physical word, 2) the reality produced by this process, 3) the illusion of the phenomenal world of separate objects. In early Vedic mythology, Maya was the power with which the gods created and maintained the physical universe.

"THAT WHICH TAKES US AWAY FROM GOD IS MAYA"

Maya in Vedic philosophy is the illusion of a limited, purely physical and mental reality in which our everyday consciousness has become entangled, a veiling of the true Self.

This illusion of profusion is created by the veiling power of the ONE Supreme Consciousness (Walter Russell's ONE MIND).

This veiling power is called Maya Shakti, or Maya.
This power creates the illusion of me and mine which creates ignorance in the individual consciousness. Those who realize this ignorance call it Avidya (a = no; vidya = knowledge: therefore ignorance or absence of knowledge). So Maya is also called Avidya. This ignorance comes to our individual consciousness through our mind: We should all strive to go beyond the mind to realize our true nature, beyond the illusion of me and mine.

The world of forms is Maya and this illusion can be seen at every level. The human body itself is not a unit of existence because of the illusion created by Maya, we see and feel our world moving and changing. We believe that our world is flowing like a river but actually it is still like a pond. When we throw a stone in it, that merely creates some ripples but eventually it is going to be still like before.

Maya is difficult to describe as there seems to be a true appearance but an untrue existence. This state of illusion is created by our minds, and all that is in the world is an illusion as well. For example; the reflection of your image in the mirror is true but not real. The fact is we are attached to an illusion of permanence. The Veiling of this illusion has to be lifted so we can see beyond this holographic illusion of the senses in this bioelectric sensory universe.

The Rg Veda and the metrics show we live in a universe which is holographic and therefore created. This is Gods mathematical, ever evolving universe of creation. The living light substance is Gods mind in action, the ever-creating ONE substance. The universe is called Prakriti and it is manifested through the vibration of the Svra, a current or life-force called the Parabrahman and the Purush. The ether is called sound and distinct vibrations cause the Pancha-Mahabhuttas, the five physical elements. These elements occur in the following order:

Akasha (Ether), Tejas (energy), Vayu (forces/fields), Pritvhi (Atomic elements) and Apas(fluid). Except for ether, all of the elements are composed of distinct indivisible particles called Paramanu (beyond atoms). I don't believe particles exist but finding a better world to explain it eludes me. Maybe Virtual Particles manifested by the ONE Mind of God.

The luminiferous ether that is acted upon by the life-giving Prana creative forces manifesting into existence the never-ending cycles of all things. The primary substance, thrown into infinitesimal whirls of tremendous velocity becomes gross matter; the force subsiding, the motion ceases and matter disappears, reverting to back to the primary substance.

The Law of Vibration, a law of nature that states that 'nothing rests; everything moves; every-thing vibrates.' The lower the vibration, the slower the vibration; the higher the vibration the faster the vibration. The difference between the manifestations of the physical, mental, emotional and spiritual result simply from different levels of vibrating energy, or frequencies.

PRANA is VRIL Magnetism

The follow information is an excerpt from VRIL OR VITAL MAGNETISM

THE SECRET DOCTRINE OF ANCIENT ATLANTIS, EGYPT. CHALDEA AND GREECE. 1911.

Vril in the human body is like electricity in an electric motor; it is *that 'which makes things "go."*

In the Arcane Teaching, the term "Vril" indicates the universal principle of vital-energy, life-force, or vital magnetism, as it is sometimes called. The term itself is believed to have had its origin in the language of ancient Atlantis, tradition holding that the Atlantean root *vri*, *m*eaning life, is the source of the word Vril, the latter expressing the idea of the vital principle or life-energy. In Hindu occultism we find the word *prana* serving a similar purpose.

"Virile" and "virility" indicate life-force or vital-energy, particularly in the sense of procreative power.

Vril is held to be a great cosmic principle of very fine energy permeating all forms of matter, and immanent in thought processes as well, being employed by the principle of mind in its work of thinking.

Vril is not identical with mind. Mind is held to be a prior manifestation of the Infinite. From the mental principle arose Vril and the grosser forms of energy, and then the forms of matter fine and gross.
In this original sense Vril is perceived to be a great universal principle from which proceeds a multitudinous manifestation of activities. Vril, in this phase of existence, cannot be defined any more than any universal principle can be defined.

The latest conception is that *Life consists in the power of independent action and movement-that* is, in the ability to act and move from inner and inherent power, and not from power or force applied from without. It is this very *power* to act and move which the Arcane Teaching holds to be the second phase of the existence of Vril. A body may possess sensation and will- ability to feel, and ability to exercise volition - and yet not be able to act and move. Feeling and will are mental states or qualities - but the power which acts and moves is something different from mind, for it is what is called vital-force, life-energy, or Vril.

Vril is the force which operates the machinery of life from the crudest movement up to the highest exercise of the brain cells of the Philosopher or mathematician.

It is by the action of Vril that the *ions,* electrons, corpuscles, or particles of elementary matter are attracted and repelled,, and by which they engage in the wild whirl around each other which resembles the movement of the planets, which attraction and repulsion and consequent U whirl" combine to form what we know as the *atom* of matter. Likewise, it is Vril which causes these atoms to be attracted and repelled, and to manifest constant vibration, thereby forming the combinations which give to us our elements of matter.

Vril, then, is the fine energy of force which enables material things to move of their own power *the power within them.* Vril *is,* in itself, this power *within,* which enables the particle or atom of matter to move to and fro; which enables the atoms to form their combinations; which causes the molecules to manifest their qualities.

Vril is not the soul, spirit, or mind, of the person any more than it is his physical body, but is a mighty natural force operating upon the body under the control of the conscious or subconscious mental faculties. Its activities manifest in and by means of the physical bodily forms and structure, it is true, but they are inspired and directed by the mind, conscious or subconscious. The physical form may and does *carry* its charge of Vril, but does not *produce* the latter. Vril *energizes and moves* the physical structure, but does not *cause* it.

In a similar manner, while Vril is active in every process of thought, it is not produced by thought; and while it energizes thought processes, it does not *produce* thought in the sense of *causing* it. Mind (in the ordinary sense); Vril, either as a principle or in its manifestation; and Matter, either as a principle or in manifestation; are *the three co-ordinate manifested principles* of the Infinite, and depend upon each other for their activities.

Not only do we see Vril manifesting in the inner movements of the *ions,* atoms, and molecules of matter, and again in its finer phases of animal and human life, but the teaching is that there are forms and manifestations of Vril so much higher than the latter that the ordinary human mind would be unable to conceive of them.
But the principle of Vril is ever the same, in high or low manifestation. Much that is called "psychic phenomena" is explainable only by knowledge of the existence, principles and laws of Vril, as set forth in the Arcane Teachings. Even the ordinary processes of thought are performed by the aid of Vril in a manner not as yet understood by ordinary men, or even by the physical scientists. It remains for the occultist to state and understand the finer forces of nature, as manifest in the processes which we call" thought."

Many of us confuse the idea of "thought" with that of "mind," but the occultist and scientist know better than this. Mind, in itself, is a great principle the exact nature of which cannot be grasped by the ordinary mind. Thought, on the contrary, is a manifestation of mind, assisted by Vril. Mind is the original cause of thought, and that it employs Vril in its thought processes just as it employs the fine matter of the brain-cells in these processes.

It is only when *the triangle of being*- Mind, Vril, and Matter is recognized, that one's full powers and energies may be manifested.

The student should bear in mind that Vril is never *manufactured* in the human body there is just so much Vril in existence - a certain amount or quantity - and this amount or quantity never can be added to, nor subtracted from, by the organism of man. The Vril so gathered, stored, and transformed is never created by the organism; nor is the Vril so used ever destroyed. The seeming creation is merely the absorption of the Vril needed, from the universal supply thereof; and the seeming destruction is merely the return of Vril to the universal supply thereof. Vril is never created nor destroyed - it merely undergoes transformation of phase, form, and use.

The mechanism of the human body involved in the absorption, storage, transfornlation, and use of Vril, is that which is known in ordinary physiology as "the nervous system." Very few persons know the facts concerning this most wonderful mechanism of the human organism, which is employed as the mechanism of the activities of Vril.

The nervous system of the human being is divided into two great systems, *viz.,* the cerebra-spinal system and the sympathetic system. The cerebro-spinal nervous system consists of that part of the general nervous system which is composed of the brain and the spinal cord, together with the nerves which emerge from the latter.

Its functions are those connected with the processes of sensation, volition, and the higher processes of thought. It conveys to seeing, smelling, hearing, and tasting. It manifests consciousness and the phenomena thereof. It attends to the functions of thought. It is the channel and mechanism of action. Through it the individual receives knowledge of the outside world, and communicates information to the outside world.

Vril is in physical manifestation in every activity or function of the body. From the slightest movement of the cell to the more complex activities of the organs of the physical body, Vril is seen to be in manifestation and activity. The subconscious planes of the mind of the individual have control of the majority of the physical activities and functions, the conscious mind not being drawn into the activity.

But in these subconscious processes Vril is ever the active force and power by means of which the work is performed. The subconscious mind without the power of Vril would be like a man without arms, hands, or tools, who would seek to perfom1 skilled manual labor.
On the other hand, Vril without the directing impulses of the subconscious mind would be like the arms, hands, and tools, apart from the directing power of the brain of the workman.

AS we have said in a preceding lesson, Vril is found in a high degree in the precise combination required for transmutation into human vital energy and nervous force in the atmospheric air which man constantly breathes into his lungs.

But science does not admit this any more than it does the existence of Vril in food and water. To science, air is merely a combination of oxygen and nitrogen with a mixture of carbureted hydrogen and carbonic acid, a trace of ammonia, and a suggestion of the four newly discovered atmospheric elements, '*viz.* : argon, crypton, metargon, and neon - or, more strictly, oxygen and nitrogen holding a mixture of several other substances in small proportions. But the occult teachings have always held that in the atmospheric air of the earth there is to be found Vril in a high degree of potency, and in a condition which renders it very easily absorbed and assimilated by the nervous system of plants, animals, and human beings. It would seem that the special combination of oxygen and nitrogen produces a condition in which the element of Vril is easily liberated under certain conditions, in such form as to be easily transmuted and absorbed.

It is a fact acknowledged by physiologists that persons who breathe through the mouth are not nearly as healthy as those who habitually breathe through the nostrils. This fact is known to the savage races, many of whom take great care in forcing their infants to acquire the habit of nostril-breathing and to avoid mouth-breathing. Occultists who practice breathing methods for the purpose of the absorption of Vril frequently moisten each nostril before beginning their exercises.
This plan is held to increase the power of the nerves of the nasal channel, and to increase the sense of smell as well. Some of the Oriental occultists draw water by suction up through the nasal passages, allowing it to escape through the mouth by means of the canal connecting the nose with the throat. This plan, by the way, is said to be a preventive of nasal disorders such as catarrh.

The nostrils should always be kept dear of obstructions, and a healthy condition preserved. Another fact known to the ancient occultists which also is unknown to modern physiology is that *the individual may largely influence the power of absorption of Vril by the action of the mind, in the form of ideation and use of will-power-* that is to say, by the familiar process of visualization or the forming of a mental picture, backed up by the use of the will. To those who may be skeptical as to the effect of the mind over a physical function of this kind, we would say that in the first place the absorption of Vril is somewhat different from the ordinary physiological function, and in fact may be considered rather as a psychophysiological process than a purely physiological one. Vril is not a material substance, but a form of energy of a very subtle nature, filling a space in the scale between matter and mind, and being in a way associated with both.

FULL BREATHING: The best authorities agree that the best possible form of breathing is that which *is based on abdominal breathing, but which also includes the filling of the middle and upper part of the lungs as well.* By what may seem to be a striking coincidence, it is noted that this particular form of breathing is that which was taught by the ancient arcane teachers to their students as a means increasing the absorption of Vril. But the coincidence is quite a natural one and it would be indeed strange had it not occurred.

For this "full breathing" method is the true, natural, normal method of breathing which natural man instinctively employs. It not only fills every part of the lungs, and exercises every part of the chest - not only secures the greatest possible amount of oxygen and Vril- but also obtains the greatest returns from the least comparative effort. In full breathing, all of the respiratory muscles are called into play; the entire area of the lungs is used; the entire machinery of the respiratory organism is exercised, strengthened and developed. There is every evidence that this and this alone, is nature's normal method of breathing. Moreover, it is known that the hardiest races of men have practiced this form of breathing.

We know this from the modern instances, and because the statuary of ancient Greece shows that muscular development of the abdomen and chest which comes only from this form of breathing. It is the first word of nature to man regarding breathing- it is the last word of science to man on the same subject. It is the best natural method - it is the best scientific method. Full breathing is not an artificial system or method of breathing but is rather a return to natural normal methods and habits. But, nevertheless, it will require some practice on the part of many students hereof, by reason of the fact that they have lost their natural instinct in the matter, and are under the dominion of the "second nature" of false habit.

EXERCISE: The following exercise will serve to develop the full breath, if conscientiously practiced.

(I) Standing erect, or sitting in a natural position, inhale slowly through the nostrils, and according to the method of " abdominal breathing" fill the lower lungs, press down the diaphragm, and push out the abdomen in front and at the sides then in a continuous effort (2) fin the middle part of the chest and lungs, as in intercostal breathing, pressing outward the mid-ribs, breast-bone and chest; then in the same continuous effort (3) fill the upper portion of the lungs, as in clavicular breathing, lifting the upper portion of the chest, slightly raising the collar-bone, slightly drawing the abdomen and thus raising the diaphragm, as heretofore explained.

The secret of mental alchemy may be stated as consisting first, last and always of the art of mental imaging, reinforced by the will. Mental alchemy, under whatever name it may masquerade, may be found to consist, at the last, of simply the power to create strong, clear, mental images, and to project them into the outer world by means of the concentrated 'Will.

"Oh, Neophyte, in the Centre of Life shalt thou indeed poise and power. The Heart of the Storm shalt thou find peace. He, who finds the centre of himself, finds the centre of the Cosmos. For at last they are ONE!" End of excerpt. There is a lot more information that people should learn about Prana and how it can benefit their lives and health.

AETHER

Its existence was really credited in the first instance as the result of the celebrated undulatory theory of light enunciated in 1801 by Thomas Young; indeed, the only function which the Ether was at first supposed to perform was to serve as the medium for the transmission of light-waves; its only activity was to undulate. Gradually, however, the important part which the Ether plays in all physical phenomena was recognised. Its connection with electric and magnetic phenomena was clearly demonstrated; and, indeed, the history of the Science of the nineteenth century may be said to be the history of the discovery of the all-importance of the Ether. The Ether literally ensouls physical matter; no physical phenomena whatsoever can take place without its agency.
Professor Sir J. J. Thomson tells us that " All mass is mass of the ether ; all momentum, momentum of the ether ; all kinetic energy, kinetic energy of the ether; " in other words, every physical phenomenon whatsoever, including matter itself, has its origin and source in the Ether.

As far back as 1882 Sir Oliver Lodge wrote of the Ether and its functions as follows : " One continuous substance filling all space : which can vibrate as light ; which can be sheared into positive and negative electricity ; which in whirls constitutes matter; and which transmits by continuity, and not by impact, every action and reaction of which matter is capable. This is the modern view of the Ether and its functions."

Energy is always found in association with matter, so that matter has sometimes been termed the Vehicle of Energy, therefore, we find energy of any kind or sort, there we find matter also, as the two are inseparably connected together. Thus, wherever we have heat, we have matter in a particular state of motion, generally understood as vibratory motion. Wherever we have light, which is also a form of energy, we also have matter in motion, that is the Aether, in a state of periodic wave-motion; and wherever we have electricity, we have again matter possibly in a state of rotatory motion. Energy, therefore, is the power which a body possesses to do work.

If, however, vibrating Aether forms structure but it is not atomic unless something gives it structure, the principle that Aether is atomic is debatable. If we could see the aether vibrating to form an atom that atom is not vibrating in the direction of propagation, but across the line in which the wave is travelling.
Thus the vibration of the air is said to be longitudinal, but the vibrations of the Aether are transversal.

Because Aether surrounds all bodies in the universe, from the smallest atom to the largest sun or star. Our sun, the source of all light, will be surrounded by what are practically spherical aetherial envelopes or shells which decrease in density as they recede from the sun.

Thus, when a wave motion is set up in the Aether around the sun by the intense atomic activity of that incandescent body, each atom of that aetherial spherical shell or envelope participates in the motion or impulse received, at one and the same time, so that the wave is transmitted from envelope to envelope, by the elasticity of the aetherial atoms which compose the envelope or shell.

Thus the light wave is always spherical in form. Further, the wave front always takes the form of a sphere, as the waves are radiated out from the luminous body in all directions, and we shall learn that the vibrations are always in the wave front, that is, take place on the surface of each of these envelopes, and these vibrations are also transverse to the propagation of the wave.

We know that light is due to a periodic wave motion of the Aether thus in the phenomena of light and heat. Aether is the medium in which the energy of light is stored, and by which it is transmitted in its passage from a luminous body, as the sun, until it comes into contact with a planet or satellite from which it is reflected, thus giving rise to light and heat.

Electricity is due to certain motions of the universal Aether but surrounds all particles and atoms of all Matter. Is Aether Electricity, or, in other words, are Aether and Electricity one and the same? We know that electricity and magnetism have an aetherial basis, and are also due to certain kinds of motion in the Aether. We know Aether has a thermal or heat basis, or a luminiferous or light basis but is Aether and electricity one and the same thing.

Electricity, whether it is electrostatic or current electricity, or electro-magnetism, are due to certain forms of motion of the universal Aether, in the same way that light and heat are also particular forms of motion of the same medium. I believe that electricity is created when magnetism is trying to restore balance and as that Aether is magnetism, Aether is therefore electricity.

"The Aetherless basis of physical theory may have reached the end of its capabilities and we see in the aether A NEW HOPE FOR THE FUTURE."
Prof. P.A.M Dirac, 1954 Nobel Prize in Physics 1933.

Mainstream Science has gone from knowing all about Aether to totally disregarding the existence of Aether and now needing Aether again to explain their theories. There has been so much time wasted it is like the blind leading the blind just stumbling around in the dark. Meanwhile occult societies haven known all along the true nature of the universe and have developed technology in secret for their own advancement.

From the Bhagavad-gita (As It IS).

"Earth, water, fire, air, ether, mind, intelligence and false ego – altogether these eight comprise My separate material energies."

The science of God analyses the constitutional position of God and His diverse energies. Material nature is call prakrti, or the energy of the Lord in His different purusa incarnations (expansions) as described in the Svatvata Tantra:

"For material creation, Lord Krsna's plenary expansion assumes three Vishnu's. The first one, Maha-Visnu, creates the total material energy, known as mahat-tattva. The second, Garbhodakasayi Visnu, enters into all the universe to create diversities in each of them. The third, Ksirodakasayi Visnu, is diffused as the all-pervading Supersoul in all the universe and is known as Paramatma, who is present even within the atoms. Anyone who knows the three Vishnu's can liberate from material entanglement."

Everything that exists is a product of matter and spirit. Spirit is the basic field of creation, and matter is created by spirit. Spirit is not created at a certain stage of material development. Rather, this material body is developed because spirit is present within; a child grows gradually to boyhood and then to manhood because of that superior energy, spirit soul, being present. Similarly, the entire cosmic manifestation of the gigantic universal form, are originally two energies of the Lord, and consequently the Lord is the original cause of everything.

Sound does not travel in waves like a stone thrown into a pond; sound is expanding spheres that create the ripple on the surface giving the illusion of waves. Beyond the illusion of the wave, there is only information, the code of God's creation.

Kepler's "The Secrets of the Universe" and "The Harmony of the World" show the tones of the planets and the harmonious unification of what many call reality. The reality of it all is that music is the sound of creation and the words forming God's matrix.
"Whenever the two forces of contraction and expansion meet and are held in some proportional balance, a being arises – and a tone is sounded. Every being is both a number and a tone, quality and quantity, both existence and value. All have the same root: the originating 1:1 (Fundamental) tone that represents …the Creator."
(Godwin, 1987 p. 191)

The vibrations of sounds or musical tones are continuously forming the manifested universe of ongoing creation. Light/Aether vibrates at different frequencies to create the gases through to solid structures of matter we call reality. The illusion of reality or consciousness is a generated illusion. The manifested ONE light makes all of us ONE. We are all in God's light but we are all individuals and can achieve Krsna consciousness but we cannot be Krsna, we cannot be god or ONE with god.

The Following information is from the **QUANTUM AETHER DYNAMICS INSTITUTE** and their PDF "Secrets of the Aether."
Concerning the Investigation of the State of Aether in Magnetic Fields: When the electric current comes into being, it immediately sets the surrounding aether in some kind of instantaneous motion, the nature of which has still not been exactly determined. In spite of the continuation of the cause of this motion, namely the electric current, the motion ceases, but the aether remains in a potential state and produces a magnetic field. That the magnetic field is a potential state [of the aether] is shown by the [existence of a] permanent magnet, since the principle of conservation of energy excludes the possibility of a state of motion in this case. The motion of the aether, which is caused by an electric current, will continue until the acting [electro-] motive forces are compensated by the equivalent passive forces which arise from the deformation caused by the motion of the aether itself.

Re-discovering the Aether

What if the ancient and universal idea of Aether proved to be the true foundation of reality? Acknowledgment of the Aether solves many problems in physics. A dynamic Aether would explain some of the most complex difficulties in the Standard Model. Imagine that the universe is an ocean of living energy. As the search for the true nature of space-time gains momentum, we are seeing that new discoveries and theories in space-time look more and more like the ancient concept of the Aether. Instead of space being emptiness, a void of nothingness, it begins to appear that space is the mother of everything. Their **Unified Force Theory** is an expansion on what has been known for a long time.
As long as the knowledge gets out about Aether that is what is important. It is hard to not believe that there has been some type of a ruse going on to mislead modern scientist about Aether. Aether science or Occult Aether Physics is suppressed knowledge and more will be explained as you read further.

Thomas Moray called Aether a sea of energy and John Keely said "The luminiferous ether - the compound interetheric element - in other words, celestial mind force - is the substance of which all visible and invisible things are composed. The fundamental mode of vibration changes as we reach the fifth subdivision [of matter], to the dominant, the diatonic third of the mass chord, which controls the vibratory states of both the etheron and interetheron.

The awful might concealed in the depths of the etheric and interetheric subdivisions utterly transcends anything Science has ever known. Even the theoretical energy value of radium now accepted by Science, pales into insignificance in comparison to the energy value of an equal amount of water subdivided to the etheric or interetheric state."

Nikola Tesla believes everything in the universe derived its energy from external sources only the existence of a field of force can account for the motions of the bodies as observed, and its assumption dispenses with space curvature. All literature on this subject is futile and destined to oblivion. So are all attempts to explain the workings of the universe without recognizing the existence of the Aether and the indispensable function it plays in the phenomena.
My second discovery was of a physical truth of the greatest importance. As I have searched the entire Scientific records in more than a half dozen languages for a long time without finding the least anticipation, I consider myself the original discoverer of this truth, which can be expressed by the statement: There is no energy in matter other than that received from the environment." Nikola Tesla

Sympathetic Vibrational Physics is all part of Vedic Physics

Sympathetic resonance or sympathetic vibration is a harmonic phenomenon wherein a formerly passive string or vibratory body responds to external vibrations to which it has a harmonic likeness.

The classic example is demonstrated with two similar tuning forks of which one is mounted on a wooden box. If the other one is struck and then placed on the box, then muted, the un-struck mounted fork will be heard. In similar fashion, strings will respond to the external vibrations of a tuning fork when sufficient harmonic relations exist between the respective vibratory modes. A unison or octave will provoke the largest response as there is maximum likeness in vibratory motion. Other links through shared resonances occur at the fifth and, though with much less effect, at the major third. The principle of sympathetic resonance has been applied in musical instruments from many cultures and times.

From Wikipedia
As everyone knows, the aether played a great part in the physics of the nineteenth century; but in the first decade of the twentieth, chiefly as a result of the failure of attempts to observe the earth's motion relative to the aether, and the acceptance of the principle that such attempts must always fail, the word 'aether' fell out of favour, and it became customary to refer to the interplanetary spaces as 'vacuous'; the vacuum being conceived as mere emptiness, having no properties except that of propagating electromagnetic waves. But with the development of quantum electro-dynamics, the vacuum has come to be regarded as the seat of the 'zero-point' oscillations of the electromagnetic field, of the 'zero-point' fluctuations of electric charge and current, and of a 'polarisation' corresponding to a dielectric constant different from unity. It seems absurd to retain the name 'vacuum' for an entity so rich in physical properties, and the historical word 'aether' may fitly be retained.

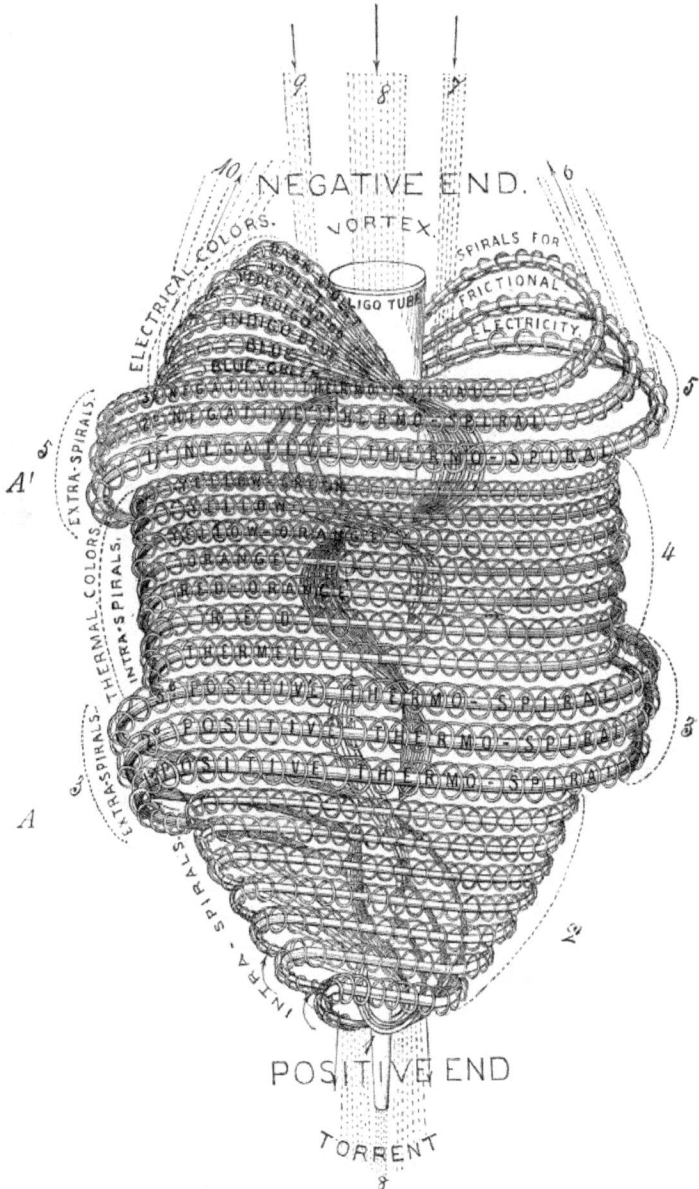

Fig. 135. The general Form of an Atom, including the spirals and 1st Spirillæ, together with influx and efflux ethers, represented by dots, which pass through these spirillæ. The 2d and 3d spirillæ with their still finer ethers are not shown.

This image is from The Principles of Light and Color by Edwin D Babbitt 1878.

The Law of Unity is universal through all matter and mind, and is the expression of wholeness, , centralization and organization.

The following two images are from "The Rational Non-Mystical Cosmos; the Mysticism of Science Exploded." Gillette, George Francis. 1933.

INTERNAL STRUCTURE OF MASS
Four Planes of Size: Moon, Earth-Moon, Solar System, Supra Solar System
Paths of Series of Four Cycles of Co-Screwed Radiators

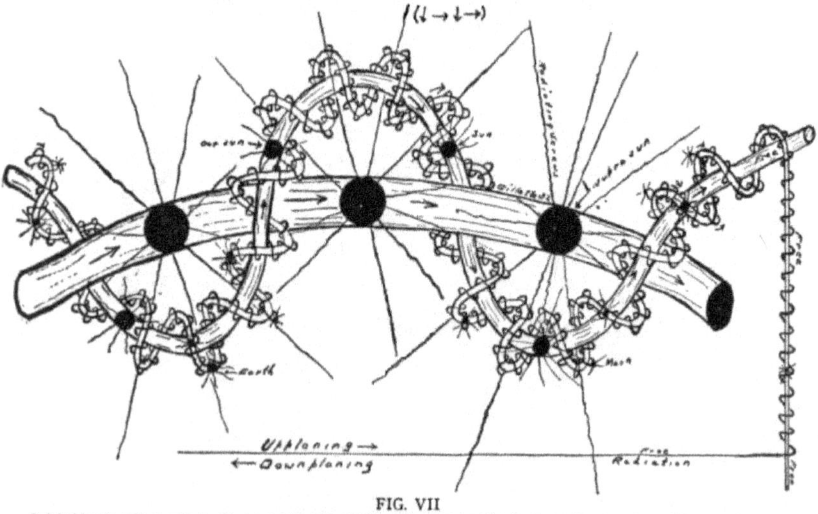

FIG. VII

Seried ultimotic coils, of coils of coils, temporarily held coiled in each coil separately, by that coil's own solar radial gravitational units of subplanely similar structure, in cosmic cycles of gravitation of descending sized units of radiation.
 The drawing represents one subunit spiralling, respiralling and again spiralling through three planes of size. It might represent one unimote spiralling upplane through electron, atom, and molecule or moon, earth and solar system. Each new larger curve indicates a higher planar coil, "re-creating" a new higherplane mass unit, with the cooperation of billions of its brother units. A first attempt was made to coil the finest wire around a larger one, draw coil off empty, recoil it, etc. But an empty coil would not "take a set" in a recoil. New core wires were found necessary to hold "set" each new recoil. And so it is in nature. Each new higherplane needs a new sun. Our great mass units in higherplania, if compacted solid, would crush of their weight, their centripetal gravitation. Great integrated mass units can build and maintain themselves only as co-swirls of co-swirls, ever of lighter specific gravity upwardly. Our mighty sun is relatively but foam upon a stein of subplane beer. The drawing shows the spiral orbital paths, each plane normal and gyrocoupled to its lower and higher brother plane. Any change of direction of topplane whips downplane counterrocking subunits' plane of rotation. Disintegration, explosion, straightens out a top spiral. Re-creation, upplane co-swirling, adds a new sun and respirals lowerplane spirals. Note that even the highest coils are made of only the finest wire—the solar cores being similarly built. (See chapter on Solar Cores.)

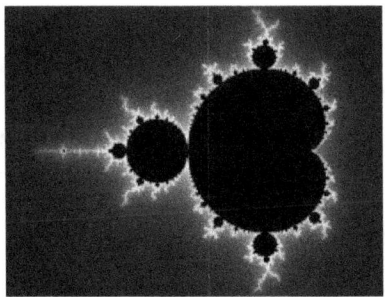

To me the below image this looks like fractal geometry. Supposedly the first pictures of this fractal were drawn in 1978 by Robert W. Brooks but the above image is from Gillette, George Francis. 1933.

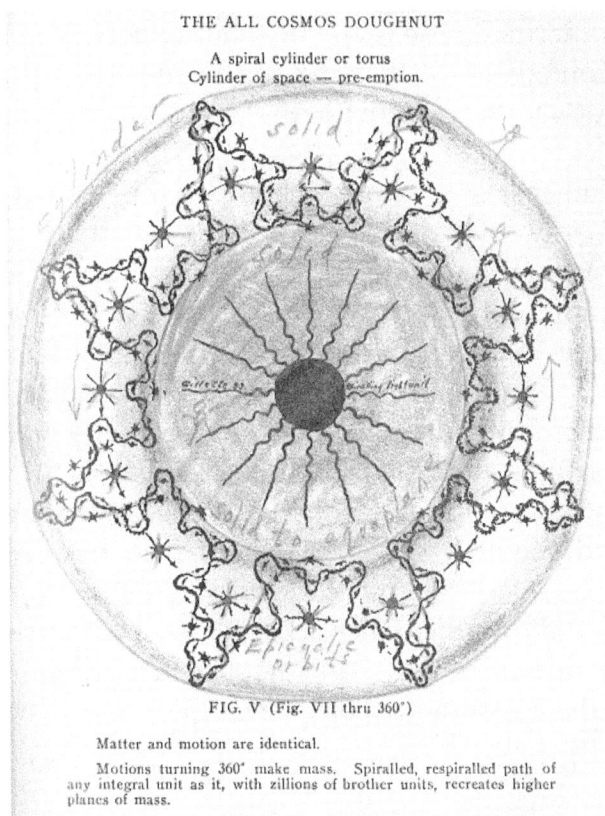

FIG. V (Fig. VII thru 360°)

Matter and motion are identical.

Motions turning 360° make mass. Spiralled, respiralled path of any integral unit as it, with zillions of brother units, recreates higher planes of mass.

In each plane, the single sun is repeated in successive positions.

This paper was awarded the UICEE diamond award (joint first grade with one other paper) by popular vote of Conference participants for the most significant contribution to the field of engineering education. Education, held in Bangkok, Thailand, from 5 to 9 July 2004.

Digital signal processing (DSP) is the technology that is omnipresent in almost every engineering discipline. It is also the fastest growing technology this century and, therefore, it poses tremendous challenges to the engineering Education, held in Bangkok, Thailand, from 5 to 9 July 2004. This paper was awarded the UICEE diamond award (joint first grade with one other paper) by popular vote of Conference participants for the most significant contribution to the field of engineering education.

It is also the fastest growing technology this century and, therefore, it poses tremendous challenges to the engineering community. Faster additions and multiplications are of extreme importance in DSP for convolution, discrete Fourier transforms digital filters, etc. The core computing process is always a multiplication routine; therefore, DSP engineers are constantly looking for new algorithms and hardware to implement them. Vedic mathematics is the name given to the ancient system of mathematics, which was rediscovered, from the Vedas between 1911 and 1918 by Sri Bharati Krishna Tirthaji. The whole of Vedic mathematics is based on 16 sutras (word formulae) and manifests a unified structure of mathematics. As such, the methods are complementary, direct and easy.

The authors highlight the use of multiplication process based on Vedic algorithms and its implementations on 8085 and 8086 microprocessors, resulting in appreciable savings in processing time. The exploration of Vedic algorithms in the DSP domain may prove to be extremely advantageous. Engineering institutions now seek to incorporate research-based studies in Vedic mathematics for its applications in various engineering processes. Further research prospects may include the design and development of a Vedic DSP chip using VLSI technology.

INTRODUCTION

Vedic mathematics is the name given to the ancient system of mathematics, or, to be precise, a unique technique of calculations based on simple rules and principles with which any mathematical problem can be solved – be it arithmetic, algebra, geometry or trigonometry. The system is based on 16 Vedic sutras or aphorisms, which are actually word formulae describing natural ways of solving a whole range of mathematical problems. Vedic mathematics was rediscovered from the ancient Indian scriptures between 1911 and 1918 by Sri Bharati Krishna Tirthaji (1884-1960), a scholar of Sanskrit, mathematics, history and philosophy. He studied these ancient texts for years and, after careful investigation, was able to reconstruct a series of mathematical formulae called sutras.

Bharati Krishna Tirthaji, who was also the former Shankaracharya (major religious leader) of Puri, India, delved into the ancient Vedic texts and established the techniques of this system in his pioneering work, Vedic Mathematics (1965), which is considered the starting point for all work on Vedic mathematics.

Vedic mathematics was immediately hailed as a new alternative system of mathematics when a copy of the book reached London in the late 1960s. Some British mathematicians, including Kenneth Williams, Andrew Nicholas and Jeremy Pickles, took interest in this new system.

They extended the introductory material of Bharati Krishna's book, and delivered lectures on it in London. In 1981, this was collated into a book entitled Introductory Lectures on Vedic Mathematics [2]. A few successive trips to India by Andrew Nicholas between 1981 and 1987 renewed interest in Vedic mathematics, and scholars and teachers in India started taking it seriously.

According to Mahesh Yogi, The sutras of Vedic Mathematics are the software for the cosmic computer that runs this universe. A great deal of research is also being carried out on how to develop more powerful and easy applications of the Vedic sutras in geometry, calculus and computing.

Conventional mathematics is an integral part of engineering education since most engineering system designs are based on various mathematical approaches. All the leading manufacturers of microprocessors have developed their architectures to be suitable for conventional binary arithmetic methods. The need for faster processing speed is continuously driving major improvements in processor technologies, as well as the search for new algorithms. The Vedic mathematics approach is totally different and considered very close to the way a human mind works. A large amount of work has so far been done in understanding various methodologies (sutras).

However, hardly any meaningful applications of Vedic algorithms have been thought of. In this article, the authors show how a successful attempt has been made to present two and three-digit multiplication operations and the implementation of these using both conventional, as well as Vedic, mathematical methods in 8085/8086 microprocessor assembling language. The authors also highlight a comparative study of both approaches interims of processing times (T states).

COMPARATIVE STUDY OF PROCESSING TIMES OF CONVENTIONAL AND VEDIC MULTIPLICATIONS FOR 8085/8086 MICROPROCESSORS

By using the above-mentioned Vedic methods for two- and three-digit multiplications, assembly programs on 8085/8086 microprocessors were written, along with the number of clock states (T states) per instruction, in order to evaluate the total processing time for each of the methods.

Similar exercises have been completed for two and three-digit multiplications utilising conventional mathematics methods. Both types of programs were successfully run to obtain the correct results (the assembly program listings could not be reproduced in this article due to paucity of space). A 3 MHz clock used in 8085 Microprocessor gives one T state equal to 0.33 microseconds. Therefore, the total processing time is the product of total number of T states and the time period of one T state (i.e. 0.33 microseconds). A comparison of the processing times for Vedic and conventional mathematical methods in the case of two- and three-digit multiplications reveal the details listed below.

Two-digit multiplication yields the results shown in Table 1. As evidenced from Table 1, a time saving of approximately 59% can be achieved using the Vedic method. Three-digit multiplication gives the results shown in Table 2. In the case of three-digit multiplication, approximately 42% of the processing time is saved. Similar results can be obtained on other processors as well. The above results are extremely encouraging so far as applications in digital signal processing (DSP) are concerned.

Most of the important DSP algorithms, such as convolution, discrete Fourier transforms, fast Fourier transforms, digital filters, etc., incorporate multiply-accumulate computations [6]. Since the multiplication time is generally far greater than the addition time, the total processing time for any DSP algorithm primarily depends upon the number of multiplications.

REFERENCES
1. Jagadguru Swami Sri Bharati Krisna Tirthaji Maharaja, Vedic Mathematics: Sixteen Simple Mathematical Formulae from the Veda. Delhi (1965).
2. Williams, K., Discover Vedic Mathematics. Skelmersdale: Inspiration Books (1984).
3. VedicMaths.Org (2004), http://www.vedicmaths.org
4. Brey, B.B., The Intel Microprocessors 8086/8088, 80286, 80386, 80486 (6th edn). Englewood Cliffs: Prentice-Hall (2003).
5. Abel, P., Assembly Language Programming (5th edn). Englewood Cliffs: Prentice Hall (2002)

The Vedic Metric, the Lotus Flower Measuring System

The 1080 Metric is The Measuring System of the Gods.

The following information is what I used to discover the Vedic Metric and bring it to light in a way anyone could understand it. All quantum metrics are derived from the Rg Veda and there is no question about that. . As you read the below passage you can realise how difficult deciphering this information was. People need to be aware and focus their thinking on the truth not just believe the lies of the establishment.

Vedic Metric conversion from the ManuScript10-ETIGammaNeT.

The Rg Veda, was the source of the I-Ching DNA code. The Quantum Phonon is conveyed through what is called K-Space (Complex Space), which itself is based on a 16 unit meter, like AUM's 16 petalled Lotus, and the 16 Purusah verses, that forms . of the Purusah universal crucifix or cross game board, the Astipada of 4 x 16 = 64 divine syllable dancers. The Ku's Muse or VakAUM Ku/Cow's K-Space is a form of Vector Imaginary Space, or the Logos "I" Magi Nation, the Purusaratha Cosmic City, from which stems the New Jerusalem of the Cosmic Christ.

The precision, with which the Manu Purusa hymn was selected and positioned, appears to have a treasure trove of bounty to its design, as this chapter does show… In the Vedic traditions, the number 90 figures directly in the dimensions of the sacred mahavedi altar ground where the special Soma rituals were conducted. Mahavedi was ascribed using precise sacred measure, in the form of an isosceles trapezoid whose foundational bases were appropriated the measure 24 and 30 with a width 36.

Thereby grafting the Moons three cycles of 8 lunar tides that span one day and night, to the 24 syllables of the tripled footed Gayatri meter, of 8 syllable per pada — 888. The base component of 30 grafted to a Pankti meter of 30 syllables to personify the 30 Muhurta hours of the day sum of these numbers is 90, which was chosen since it represents one-fourth of the year. The sum represents an example of equivalence by number, the shape of a trapezoid with its specific dimensions, was chosen, since this shape generates many Pythagorean triples. $90 \times 4 = 360$, they Lunar calendar of the Rg Veda, of 360×30 Muhurtas $= 10,800$ Muhartas. The number of Stanas in the Rg Veda, and a direct play on the 108 solar circumferences distance to the Earth from the Sun, and the 108 lunar circumferences distance from the Moon to the Earth, as ONE Single Skambh Pillar… Aja Eka Skambh, of the Purusah.

Purusah, as a unique appearance in one sole hymn of the Rg Veda, like a unicorn, thereby has some cross sybiosis to that of the Ekapada Viraj meter: which is a Viraj meter, but consisting of a single Pada, of 10 syllables. Thustranscribing the 10 Omni-stretching fingers of the Purusa.

The heaven's altar is given 261 stones, which has its relationship in the Purusah Sukta — it being the 90th hymn, of 163 words, being the only occurrence of the name Purusa, which occurs 8 x renders:
90 + 163 + 8 = 261, this thus could be revealing that the 3/4's of the Manu Purusa that ascended beyond by 10 fingers, is represented by the number 261, the altar of heaven, of the full circle, 360°, of the triplet reality, leaving the 21 stones of the Earth altar and the 78 of the sky altar... Hence, summing to the number 99, which the Veda in almost all instances, decrease to be associated to the sleep demons, such as those with the 99 fortresses of Sambhara, that guard the Soma Cup, at the galactic centre, or Visnu Nabhi When Purusah is seen as hymn 90 of the mandala 10, as the 10 fingers, then Purusah as 90 + 10 = 100. Signalling back by 10 to the 1000 heads of Purusah. 100 Spirits or breath of life is called Atman or Atamanu, the thought atom associated to Purusah, and is
the life-breath, from which atum, atom, and German atem, breath. 10 fingers, 10 + 90 = 100 x 10.

10 x 100 = 1000. One quarter of this is 25, which is associated to the letter "M" of AUM, being the 25th Sanskrit consonant, and the 5 elements x 5 senses as the 25 Tattva's, also evident in the Vedic atomic theory's is 75, which is the precise figure one gets when one divides 3 by 4, as the 34 vowels = 0.75.
Symbiotic to this, when 360 days of the Moon through the Lunar 27 Meru mansions of one year, 1/4 of this is 90, and is 270, perfect decimal of the Meru 27, the AUM mountain of creation, which is also the game board of Purusa's body. 270 days are also the 9 months of gestation. The Meru 9 triangles are 3 x 3 x 3 = 27 lines.

The 27 lines on the general cubic surface that are raised to the hypercube by the decimal 10 (Reiman metric tensor), mapping a 27 dimensional algebra.

In addition, with Purusah being the body of the Rg Veda, of 432,000 syllables, then 16 divisions renders 27,000. This is also symbiotic to Purusah as the sacrifice being spread and bound by the gods, "for him there were enclosing six, and thrice seven." (10.90.15) Since 21 (3 x 7) plus 7 = 28, then one has the 27 Meru lines plus the Bindu point, which in the Purusah game is his golden navel from "which the middle realm of space [akasa], arose." This navel is the Spiritual Sun/Moon as one, and the 28 solar mansions are thus melded to the Lunar ones (27), just as the sun's equator rotates 360° in 28 days, and the Moon's cycle of 27 days of the sidereal month. The meter called Pura-usnih: 28 syllables made by three Padas (or 3 x .), one holding 12 syllables, another 8 , the last 8 syllables = 28.

Chandaspurusa "the body of the meters" conveys the meters as the building bricks of the pyramid hymn altars that self-embed to build larger manifolds of dimensional realities, indexing plural layers of self-referenced structure (poetry, musical harmonics, algebra, bioelectronics, astronomy, botany, neocybernetics, architecture, living wisdom, numinous revelation)

In fact the Vedic literature vividly describes the NADA and Nadi lines from suns, but more importantly both the NADA Virtual Sound (Logos Cow Milk flow and star horns) and phonon resonance grids (SARA "weave grid" RASA Essence River) — i.e. the black body sound particle, phonon's through ultra-high ordered water crystals, in the zero temperatures of space. In a superconductive arrangement, with the plasma forming around the superconducting ring currents, and producing gamma rays, just as has been observed in the M-state superconducting like platinum group elements in a high ward spin state, with the gamma weird radioactivity emanations.

1080 and the significance of 42. My moment of enlightenment when I discovered the truth uncovering the lies taught as Quantum Physics

I refer to the article, Calendar Configurations, by Rabbi Moishe Kimelman, on page 85/86 of the April 20, 2001 issue of **Hamodia**, for those interested in a summary explanation respecting the numerous calculations, formulas and numbers that we hold as a tradition from the time of Moses. Rabbi Kimelman cites the Gaon of Vilna, who notes that the number 793 came before the number 1080. The hour could have been divided simply into 15 portions, of which 11/15 is equal to 792 parts of an hour. Instead we are forced to use the large number, 1080, so as to permit the more precise 793. The number 793 is important not only for measurement accuracy but also, other, more mysterious reasons. Thus, Rabbi Avrohom Chaim Carmel points out the following. Multiply 29 days by 24 to get hours. Add the additional 12 hours and multiply by 1080 to get parts of an hour. Finally, add 793 to the result to obtain all the parts of an hour. This yields the descending order 765,432 + 1. This number has kabbalistic implications. But I do not know what it means.

The Omni-dimensional ancient artifact at the genius of history is the answer to life, the universe and everything. There is only one system of measuring time, distance, weight and everything else. The RG Veda and Vedic Cosmology is where you find the answer to the number 42. The Number 1080 is very important and not just for HD TV and Video.

Image from "The heavenly time machine" by Morris Engelson. As you can see there once were a calendar and system of measuring time that used 1080 and 793. I believe this was before the time of Mosses. The number 793 goes with the number 13 as a divisor this gives you 61mm inch or in this case 61 seconds in relation to time.
If you examine the extract from the Hindu Cosmological times cycles below and look for 1 precessional year = 25714 2/7 sidereal years 7 precessional years. This is all you have to do is divide 1080 by 42 to give you 25.71428571 and there you have one of the Quantum Metrics. They already knew the answer from the ancient RG Veda so all they had to do was write a complicated equation to hide the truth.

13. DEMONSTRATION OF THE CONSTANT OF PRECESSION

The caturyuga of 4320000 years is the unit of reference for determining the constant of precession used in the construction

The constant of precession is

```
50".4 = 0°.014 = 7/500 degrees of precession          per sidere
```

This is equivalent to one degree of precession in

```
71 3/7 = 71.42857... sidereal years.
```

The relevant infrastructure of the kalpa period (see Table III) is

```
1 manu               = 71.4       caturyugas 1/14 of introductory dawn
```

In the interval of 1/14 kalpa there are

```
(71 3/7)(4320000)(0°.014) = 4320000 degrees of precession
```

Other relations of interest are

```
1 precessional year     = 25714 2/7 sidereal years 7 precessional years
```

The image is form the Exegesis of Hindu Cosmological Time Cycles by Dwight William Johnson 1942-2008
http://westgatehouse.com/cycles.html

1080 and 793 are the dividends used and numbers 1 to 90 are the divisors. 1080 High definition and the number 42 makes up part of our holographic universe, it is an illusion manifested by an Omni-dimensional artifact.

It is very difficult not to judge these Quantum Physicists as lowlife scummy fraudsters with no integrity but if they did not blatantly lie about the true nature of where the information came from, they would not get any funding for research. This is the system that has been put in place, and clearly, it needs to be overhauled. To suppress the true foundation of physics and create an entire system based on lies and misinformation is just another con-job perpetrated on humanity.

436   BOOK V

system of one octave, with all the positions by means of which natural melody is conveyed in music. The sole difference is in the fact that in our harmonic divisions indeed both ways jointly start from one and the same term, whereas in the latter case in the motions of the planets what was previously ♭ now in the soft kind becomes G."

In the motions of the heavens like this:

Through the harmonic divisions like this:

For just as in Music the proportion is 2160:1800, or 6:5, so in the former system, which is expressed by the heaven, the ratio is 1728:1440, that is also as 6:5, and similarly for several other cases:⁹⁷

	2160	:1800	:1620	:1440	:1350	:1080
as	1728	:1440	:1296	:1152	:1080	:864

You will now therefore wonder no more at the establishment of the most excellent order of the sounds or steps in the musical system or scale by men, since you see that all they are doing in this respect is aping God the Creator, and as it were acting out a particular scenario for the ordering of the heavenly motions.

---

⁹⁶ In the hard scale, taking the aphelion motion of Saturn as G, the perihelion motion of Saturn becomes ♭. In the soft scale, the perihelion motion of Saturn is taken as G. We may say that the ♭ of the hard scale is transposed to become the higher G of the soft scale. The clef at the beginning of the soft scale indicates the transposed scale. If this is ignored, we see the position of the notes in relation to the hard scale.

⁹⁷ The relationship also holds for the two cases omitted from this list; that is 2160:1920 = 1728:1536 and 2160:1215 = 1728:972.

The following image is from "The Etheric Formative Forces in COSMOS, EARTH AND MAN" By Guenther Wachsmuth 1932.
Notice the relationship between the earth breathing rhythm and that of man.

CHAPTER FOURTEEN

The Number 1080

In contrast to the solar 666, the associations of the number 1080 are terrestrial and lunar, of water as opposed to fire. All these canonical numbers have correspondences in many different fields. 1080 expresses the influence of the moon on the forms and cycles of nature, and relates the spirit of intuition, or the guidance of the unconscious mind, to the moon-drawn tides in the waters of the earth. It also has a particular reference to the measurement of time and astronomical distances, illustrated in the following examples.

There are 10,800 seconds in 3 hours; 1080 is the average number of breaths which a man takes in 1 hour.

1080 years is a twenty-fourth part of the great year, so it takes 1080 × 2 years for the sun to progress through one sign of the zodiac. Heraclitus wrote that civilisation is destroyed by fire every 10,800 years, and 108,000 years is a quarter division or season of the Hindu Kali Yuga of 432,000 years.

In his *World System* Galileo wrote, 'The apparent diameter of the sun at its average distance is about half a degree of 90 minutes; this is 1,800 seconds, or 108,000 third order divisions.... The diameter of the sun contains the diameter of a star of the sixth magnitude 2160 times (1080 × 2).' Hipparchus counted 1080 stars of first magnitude brightness.

In metrology: 1080 feet = 888 remens and 1080 square megalithic yards is equal to 888 square yards. Further instances of the number 1080 in the ratios of metrology and sacred geometry are given in earlier chapters.

There are 108 beads in the Buddhist rosary, 10,800 stanzas in the *Rigveda*, each with 40 syllables, and 10,800 is the number of bricks in the Indian fire altar.

---

in the following description. When iron filings lie in a space outside the field of a distant magnet, they will not be drawn by its attractive force, but, if we move the magnet toward that space until such space lies within the magnetic field, everything in that space will come under the control of the magnetic action. *The alternating barometric pressure is due, in like manner to this; whether the earth organism sends its inward-drawing suctional etheric forces above the solid earth into the atmosphere—that is exhales; or draws these back into their open spheres, of activity—that is, inhales.*

We shall see that this great breathing process of the earth organism is always carried out in rhythm, in the alternation of day and night, so that the same thing which man performs in one breath (inhaling and exhaling) *many times each day* is performed by the earth once in 24 hours, that is, in the manner indicated in the diagram on pages 52-53. We will here first only point out that there is actually a definite cosmic relation between the breathing rhythm of the earth organism and that of each individual man. A man who breathes normally takes 18 breaths in a minute, or (18 × 60) 1,080 in an hour, or (1,080 × 24) 25,920 in a day of 24 hours. Now, this number of years that the sun requires in order to be at the time of the spring equinox once successively in every sign in the circle of the zodiac. To such cosmic rhythms, to which the earth organism is adjusted just as is the human organism, we shall later return ; for the moment we must only indicate that the breathing rhythm of the earth organism is induced by the alternating activity of the same etheric forces that control the breathing activity of man (Chap. XII). As will be illustrated in later diagrams, the breathing rhythm of the earth organism—which expresses itself, and may be proved, by atmospheric pressure, barometric level, vertical current, potential gradient, humidity of the several strata, degree of induction of the air, degree of emanation of the earth, etc. etc.—depends upon the rhythmical alternation by which the earth organism exhales the chemical ether into the light-ether zone (the atmosphere) and then draws it back into the solid earth ; that is, it depends upon the interchange between light ether and chemical ether, or, in other words, an expanding and a compressing force in the atmosphere, chiefly in the lower strata.

You will see a pattern starting to emerge. This image is from " The Harmony of the world" By Johannes Kepler first printed around the year 1619.
Notice the motions of the heavens and harmonic divisions.
This is nature's music scale.

METRE.
SUKTA = 1 HYMN
STANZA = Paragraph/Verse (usually of 4 verses or lines).
PADA= line "quarter", whose 4 feet each hold the metrical unit, like 8 syllables.

Usually a stanza paragraph is composed of pada line of the same type. The rare stanzas are formed by lines of different length, as an inter-metric. There are 16 meters:

1. gaayatree Six syllables per quarter (Class name -)
2. Usni
3. anuShTubh Eight syllables per quarter (Class name -)
4. Brhati
5. Viraj
6. triShTubh Eleven syllables per quarter (Class name -) :
7. jagatee Twelve syllables per quarter (Class name -)
8. shakvaree Fourteen syllables per quarter (Class name -)
9. aShTi Sixteen syllables per quarter (Class name - aShTi)
10. dhrta Eighteen syllables per quarter (Class name -)
11. prakrti Twenty one syllables per quarter (Class name -)
12. Pankti
13. atishakvaree Fifteen syllables per quarter (Class name -)
14. atidhrti Nineteen syllables per quarter (Class name -)
15. aakrti Twenty two syllables per quarter (Class name -)
16. atyaShTi Seventeen syllables per quarter (Class name -)

of which the first 7 meters are the fundamental. There being 90 all in all, when rare stanzas varieties are included.

6. METRE

The hymns of the RV. are without exception metrical. They contain on the average ten stanzas, generally of four verses or lines, but also of three and sometimes five. The line, which is called Pada, (quarter) and forms the metrical unit, usually consists of eight, eleven, or twelve syllables. A stanza is, as a rule, made up of lines of the same type; but some of the rarer kinds of stanza are formed by combining lines of different length. There are about fifteen metres, but only seven of these are at all common. By far the most common are the Tristubh (4 x 11 syllables), the Gayatri (3 x 8), and the Jagati (4 x 12), which together furnish two-thirds of the total number of stanzas in the RV. The Vedic metres, which are the foundation of the Classical Sanskrit metres except two, have a, quantitative rhythm in which short and long syllables alternate and, which is of a generally iambic type. It is only the rhythm of the last four or five syllables (called the cadence) of the line that is rigidly determined, and the lines of eleven and twelve syllables have a caesura as well. In their structure the Vedic metres the Aveșta shows, the principle is the number of syllables only, and) those of Classical Sanskrit, in which between the metres of the Indo-Iranian period, in which, as the Aveșta shows, the (except the sloka) the quantity of every single syllable in the line is fixed. Usually a hymn of the Rgveda consists of stanzas in the same metre throughout; a typical divergence from this rule is to mark the conclusion of a hymn with a stanza in a different metre. Some hymns are strophic in their construction. The strophes in them consist either of three stanzas (called trca) in the same simple metre, generally Gayatri, or of two stanzas in different mixed metres. The latter type of strophe is called Pragatha and is found chiefly in the eighth book.

These are two examples of the original format of Vedic metric and by all accounts difficult to understand. It is explained in terms of having a musical structure and being a system of measurement.

The Samrat Yantra, at Jaipur, designed by Jai Singh, measuring at its base 147' and 90' high calculates time within one second of accuracy every day.

The length of a meter is 1080mm and there are 4 feet to a meter making a foot value 270mm

http://www.vedarahasya.net/vedrea3.htm

The Samrat Yantra

The 90 metrics

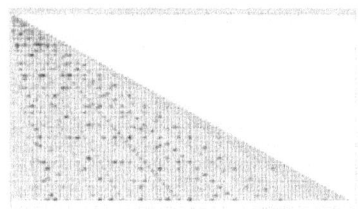

The ninety steps of the Yantra have to do with the ninety hymns of the Rg Veda which are the ninety metrics of Vedic physics the science of all oneness. If you note the progressive triangular incline of the picture above and the steps of the Yantra and a sideways view of the chart of the 90 Vedic metrics Pictured next to it. It is all the same harmonics so why would the wave not look any different?

On the other page is a screen shot of the 90 metrics. See everything is related in one way or another. Time, weight and measurement ARE ALL measured on the same scale. Now days we have may different formats for weight, temperature speed etc. but there is only one system that underlies all systems. It is all just to confuse you and keep you as far away as possible from the truth. This is a hand drawn copy of the three Metrics used below so you can envision how they all work together. Remember 1080mm long. Examples using these three metrics can be seen further in this book.

These three are 1080 / 42, 45, 48 you can see examples of each inch value

These are scaled down images of the 90 Vedic metrics that integrated into all life and all dimensions. The red arrows below indicate the 540mm or half the 1080 Vedic meter. Every second metric contains 540mm as value of one of its progressive inch values. The coloured sections are markers for the overlapping harmonics contained in each scale. To obtain the bottom image all I did was to align every 1080mm value.

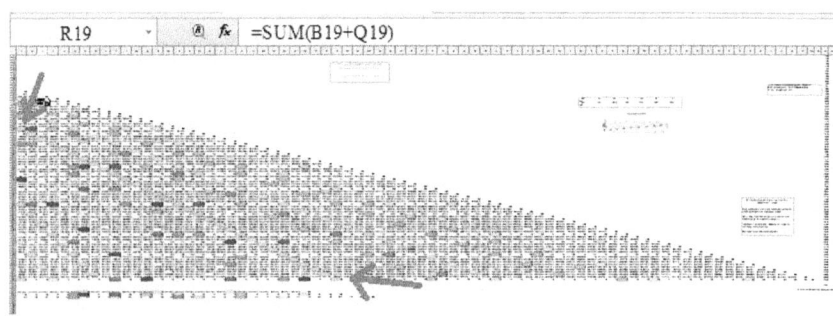

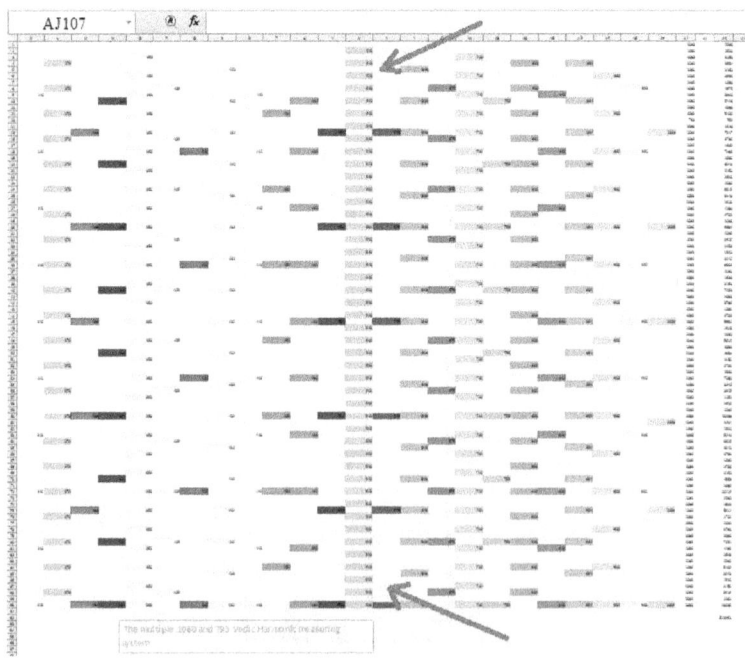

VEDIC TIME TRAVEL

Vedic Units		Modern Unit		Equivalent		Equivalent		Equivalent		Equivalent	
Paramanu		0.0000131687242798354	seconds								
Anu		0.00002633744859670	seconds	2 Paramanu		2151875 second		Primodial Atom			
1 trasarenu		0.0001580246913580250	seconds	6 Anu		4.151875 second					
1 truti		0.0004740740740740	seconds	3 trasarenu		8.50625 second					
1 vedha		0.047407407407407	seconds	100 truti		8:16875 second					
1 lava		0.142222222222220	seconds	3 Vedhas		32/675 second					
1 nimesha		0.4266666666666670	seconds	3 lavas		32/225 second					
1 kashthas		1.280	seconds	3 nimesha		32/75 second					
1 kshanas		6	seconds	5 kshanas		32/25 second					
1 laghu		96	seconds	15 kashthas		1.6 Minutes					
1 nadika		1,440	seconds	15 laghus		24 Minutes					
1 muhurata		2,880	seconds	2 nadika		48 Minutes		0.8 Hour			
1 Prahara		10,800	seconds	7 or 8 nadika		180 Minutes		3 Hour		Difference is adjusted in 13 month after 3 years - *Mal Masa*	
1 Divasa & Ratri		86,400	seconds	8 praharas or 30 muhurata		1440 Minutes		24 Hour		1 Day	
1 Paksha		1,296,000	seconds	15 complete Days		21600 Minutes		360 Hour		15 Days	
1 Masa		2,592,000	seconds	2 Pakshas		43200 Minutes		720 Hour		30 Days	
1 Ritu		5,184,000	seconds	2 Masa		86400 Minutes		1440 Hour		60 Days	
1 Ayana		15,552,000	seconds	3 Ritu		259200 Minutes		4320 Hour		180 Days	
1 Hindu Year		31,104,000	seconds	2 ayana		518400 Minutes		8640 Hour		360 Days	
1 Solar year		365.2587565	Days								
1 Pitr day		30	Days	1 masa							
1 Pitr Month		900	Days	30 Pitr Days		2.5 Years					
1 Pitr Year		10,800	Days	12 Pitr Months		30 Years					
Life of Pitr		1080000	Days	100 Pitr Years		3000 Years					
1 Deva day		360	Days	1 Hindu Year		30 Years					
1 Deva Month		10,800	Days	30 Deva Days							
1 Deva Year		129,600	Days	12 Deva Months		360 Years					
Life of Deva		12960000	Days	100 Deva Years		36000 Years					

VEDIC TIME TRAVEL

Vedic Units		Modern Unit	Equivalent	Equivalent	Equivalent	Equivalent
Paramanu	0.00001316872427983540	seconds		2/151875 second	Primodial Atom	
Anu	0.00002633744855967080	seconds	2 Paramanu	4/151875 second		
1 trasarenu	0.00015802469135802500	seconds	6 Anu	8/50625 second		
1 truti	0.00047407407407407400	seconds	3 trasarenu	8/16875 second		
1 vedha	0.04740740740740740	seconds	100 truti	32/675 second		
1 lava	0.14222222222222220	seconds	3 Vedhas	32/225 second		
1 nimesha	0.42666666666666670	seconds	3 lavas	32/75 second		
1 kshanas	1.280	seconds	3 nimesha	32/25 second		
1 kashthas	6	seconds	5 kshanas			
1 laghu	96	seconds	15 kashthas	1.6 Minutes		
1 nadika	1,440	seconds	15 laghus	24 Minutes		
1 muhurata	2,880	seconds	2 nadika	48 Minutes	0.8 Hour	
1 Prahara	10,800	seconds	7 or 8 nadika	180 Minutes	3 Hour	Difference is adjusted in 13 month after 3 years - Mal Masa
1 Divasa & Ratri	86,400	seconds	8 prahara or 30 muhurata	1440 Minutes	24 Hour	1 Day
1 Paksha	1,296,000	seconds	15 complete Days	21600 Minutes	360 Hour	15 Days
1 Masa	2,592,000	seconds	2 Pakshas	43200 Minutes	720 Hour	30 Days
1 Ritu	5,184,000	seconds	2 Masa	86400 Minutes	1440 Hour	60 Days
1 Ayana	15,552,000	seconds	3 Ritu	259200 Minutes	4320 Hour	180 Days
1 Hindu Year	31,104,000	seconds	2 ayana	518400 Minutes	8640 Hour	360 Days
1 Solar year	365.2587565	Days				
1 Pitr day	30	Days	1 masa			
1 Pitr Month	900	Days	30 Pitr Days	2.5 Years		
1 Pitr Year	10,800	Days	12 Pitr Months	30 Years		
Life of Pitr	1080000	Days	100 Pitr Years	3000 Years		
1 Deva day	360	Days	1 Hindu Year			
1 Deva Month	10,800	Days	30 Deva Days	30 Years		
1 Deva Year	129,600	Days	12 Deva Months	360 Years		
Life of Deva	12960000	Days	100 Deva Years	36000 Years		

Vedic Units	Deva Year		Hindu Years			
1 Deva Year	1		360			
1 Charan	1,200		432,000		Deva Years	
	Deva Year	Charan	Hindu Years	Dawn	Kratyuga	Twilight
Satyuga	4,800	4	1,728,000	400	4000	400
Treta Yuga	3,600	3	1,296,000	300	3000	300
Dwapar Yuga	2,400	2	864,000	200	2000	200
Kali Yuga	1,200	1	432,000	100	1000	100
1 Mahayuga	12,000	10	4,320,000			
					PRESENT	
1 Manvantara	852,000	71 Mahayuga	306,720,000	Brahma Year		51
		Earth is submerged into water		Month		1
1 sandhi kala	4,800	4 charans	1,728,000	Kalpa		1 shvetavaraha
		Period submerged		Stage	Daytime	
1 Manu	856,800		308,448,000	Manvantra		7 Vaivasvata
				Mahayuga		28
1 Kalpa	12,000,000	14 Manvantara & Sandhi Kalas & 1 adhi sandhi OR 1000 Mahayuga	4,320,000,000	Yuga	Kali	From February 18, 3102 BC
Day	12,000,000	Manifestation of Universe, Day is manifestation				
Night	12,000,000	Demanifestation of universe, Night is Brahma Sleep				
1 day of Brahma	24,000,000	2 Kalpas (Day & Night)	8,640,000,000			
1 Second of Brahma	278		100,000			
1 Minute of Brahma	16,667	60 Brahma Seconds	6,000,000			
1 Hour of Brahma	1,000,000	60 brahma minutes	360,000,000			
1 month of Brahma	720,000,000	30 Brahma Days	259,200,000,000			
1 year of brahma	8,640,000,000	12 Brahma Months	3,110,400,000,000		Prepared By: Vinay Mangal	
1 Pararddham	432,000,000,000	50 Brahma Years	155,520,000,000,000		vinaymangal@gmail.com	

These two charts are of the Hindu Time cycles to give you an idea of where the information and numbers come from.

Below is also an example of the three measuring systems of 22.5, 24 and 25.27428571 mm harmonic inch values. When you observe the harmonic relationship of the measuring system for yourself, you will understand immediately that what you are seeing is the truth.

A) 1080 ÷ 42 = 25.71428571 mm inch. This relates to space and the Hindu cosmological time cycles that proves that this system of measurement is the only system used in all cosmological calculations.

B) 1080 ÷ 45 = 24 mm inch. This represents the centre of perfect harmonics. Being the centre of the three systems is the perfect harmonic balance Phi ratio of 1.666666667 achieved two ways 360 ÷ 216 or 5 ÷ 3.

C) 1080 ÷ 48 = 22.5 mm inch. This is the system of measurement for the Ford motor companies' magneto flywheel.

D) 793 ÷ 13 = 61 mm inch the only one that is not Ten Eighty. There is proof of 793 incorporated in a musical scale in the book "Harmonic Proportion and form in nature, Art and Architecture" along with many examples of plants, animals and the human form all designed to the principles of nature and the Vedic metric. There are 4 other systems between the 42 45 48 starting points they are 1080 / 43, 44, 46 and 47.

The distance of 1800 meters is equivalent to ONE Vedic mile. The true length for ONE mile is 1080 x (5/3) or 1.666666667 = 1800 meters. One Foot is not 12 inches but 270 mm. There are four feet at 270mm per 1080 meter; not 12(25.4mm) inches per foot and about 3 feet per meter like in the English inch.

Using the BaGua, I Ching symbol arrangement I will explain how to combine 3 of the ninety metrics. The three inner segments in Bold appear to be three different lengths to represent something that looks like a wedge.

Using 3 of the metrics to explain "same length different scale" Using 9 as the length and the different metrics which are different inch values. Example 22.5mm inch value x 9 = 202.5 mm, 24mm inch value x 9 = 216 mm and 25.71428571mm inch value x 9 = 231.4287712 mm see they are all the same length of 9 inches but on a different metric scale.

You may be wondering what the I-Ching symbol has to do with Vedic physics.

I believe the ONE science of all creation has many characteristics that have be separated over time and have popped up in different cultures. This includes the swastika in different religions and magic squares which I will go into later. Each culture has its own way of interpreting the same thing.

The Computer Binary Code comes from the I-Ching.
Gottfried Wilhelm Leibniz (1646–1716) is often described as the last universalist, having contributed to virtually all fields of scholarly interest of his time, including law, history, theology, politics, engineering, geology, physics, and perhaps most importantly, philosophy, mathematics and logic [1, 7, 9].
...Leibniz was not the first to experiment with binary numbers or the general concept of a number base [5]. However, with base 2 numeration, Leibniz witnessed the confluence of several intellectual ideas of his world view and mystical ideas of order, harmony and creation, with 0 denoting nothing and 1 denoting God [13]. Additionally his 1703 paper contains a striking application of binary numeration to the ancient Chinese text of divination, the Yijing (I-Ching or Book of Changes). Surprisingly he believed that he had found an historical precedent for his binary arithmetic in the ancient Chinese lineation's or 64 hexagrams of the Yijing. This he thought, might be the origin of a universal symbolic language. - See more at: http://www.abovetopsecret.com/forum/thread227867/pg1#sth ash.n8NKtnwd.dpuf

The I-Ching & the Genetic Code.

German scientist, Martin Schönberger book The I Ching & the Genetic Code, discovered that the two Codons which contain the genetic-chemical message "to stop" have the same numeric structure of hexagram 63, After Completion, and the Codons which, so to speak, act to say "Go" on a genetic level, correspond to the opposite hexagram 64, Before Completion. In the DNA they serve as punctuation between code sequences. In the I Ching, we have hexagrams # 63 and # 64, which serve the same purpose. They both have a Duality Process. DNA is the blueprint which tells the cells how to create proteins, whereas RNA is the reverse copy of DNA which carries out DNA's instructions for protein production. This corresponds with the Yin (the yielding), which 'yields' or delivers the information and the Yang (the firm), which 'firms up' or utilizes the information and causes changes.

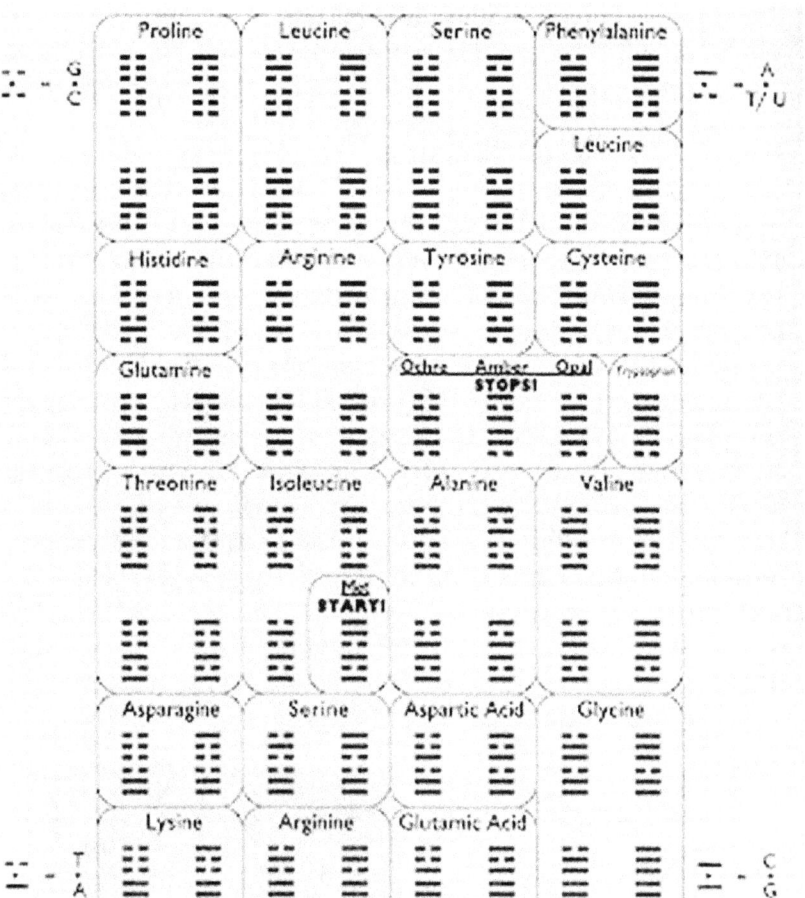

The Coral Castle

The Coral Castle Device of Ed Leedskalnin with the Vedic Metric applied to it.

For those who are unaware Ed Leedskalnin built the Coral Castle using a device that controlled the effects of gravity. Using magnetic current and different harmonic measurements you can control magnetism and therefore gravity. Ed's device was built on the flywheel and engine block of a Model T Ford.
http://coralcastle.com/

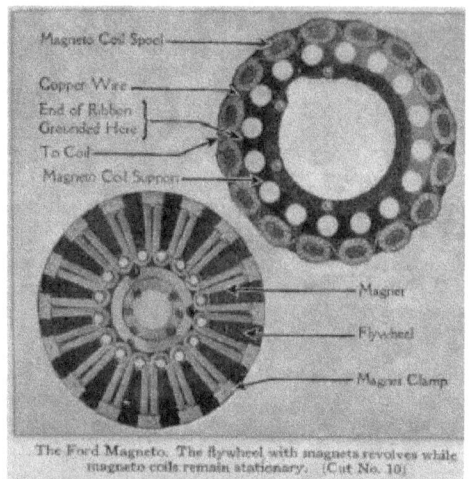

The Ford Magneto. The flywheel with magnets revolves while magneto coils remain stationary. (Cut No. 10)

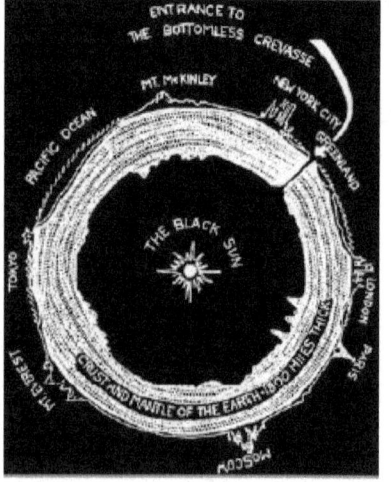

This device is built on Henry Ford's Flywheel the sweet 16. Starting with 22.5mm x 9 = 202.5mm then x 16 = 3240 (3240/22.5 mm inch) = 144 inches the original Henry Ford flywheel magneto. Left: The Black Sun symbol for the Thule Society who were behind Hitler's rise. It is a fact that the Germans had UFO's during the war and if you put in a bit of effort you will find a lot of information about it. Below is the sweet 16 as Ed calls it. The 16 v-magnets are attached to the flywheel and as it turns it produces alternating current. This object is used to control gravity.

The 16 points on the star are the 16 rays of the BLACK SUN are the pole of positive and negative energy. The C.I.A uses the same Sumerian symbology in their emblem

ideologically condemned "Jewish physics" of relativity, but quite in line with the "Aryan" physics of vorticular spin polarization, quantum mechanics, and its mathematical prediction of a vacuum energy flux, or "zero point energy." It goes without saying that the Black Sun symbolism formed a central doctrine to the pre-Nazi secret society, the Thule Society. The symbol of the Black sun was also adopted as an emblem for Von Liebenfels' New Templars.

The swastika itself, in this context, becomes not only a well-known symbol from ancient esoteric traditions, but also a talisman of ceremonial magic on a celestial scale, deliberately chosen to mirror the apparent rotation of a well-known constellation around the north pole of the earth's axis of rotation. This whole concentration on quantum mechanics, black suns, action at a distance, and celestial rotation gave a strong ideological influence to Kammler's SS think tank, for as will be seen, vorticular and non-linear physics apparently became two of its primary theoretical and experimental focuses.

The Symbol of the Black Sun Adopted by the Thule Society, Contemporary German Federal Law Forbids it to be Displayed.

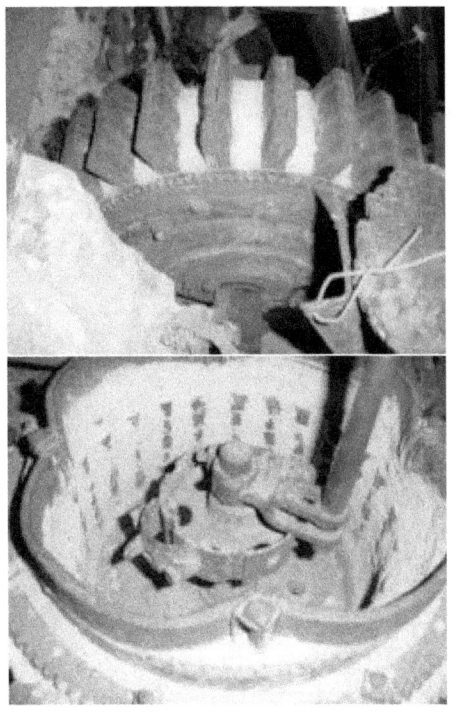

Ed's device is used to lift coral blocks weighing many tons. There are several clues in this image to how the device worked including a steel tube and the chain used as a coil. The links in the chain act like locks strengthening the magnetic current as it flows to the blocks that require lifting. The steel tube has a winding inside it of a certain tuned length as well.

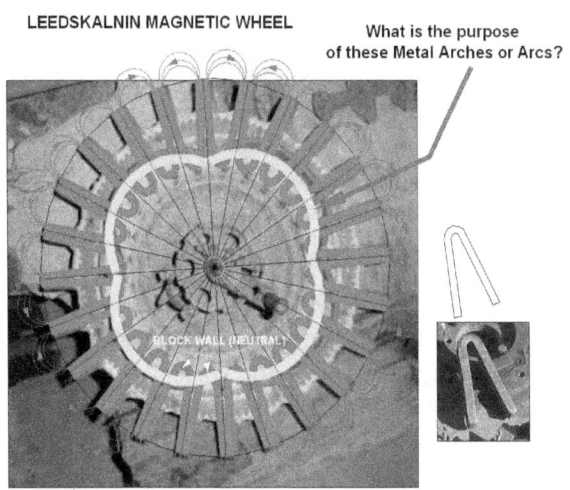

LEEDSKALNIN MAGNETIC WHEEL

What is the purpose of these Metal Arches or Arcs?

BLOCK WALL (NEUTRAL)

MAGNETIC FIELD LINES

Each magnet is 24mm inch x 9 =216mm or nine inches in the scale. The top section is 24mm inch x 9 = 216 then x (24 magnets per layer x 5 layers) 120 magnets = 25920mm. 25920 years is elliptical wobble of the earth. The two sections of 25920/3240= 8 or one octave. These are of the specifications of the Coral castle device.

Flux Pattern of a 4 pole 3 phase 48 section electric motor.

The 24 U-magnets together create 48 lines of magnetism just as you see protruding from the Star of David. I DO NOT think it is a coincidence that the secret to Ed's device appears in the Grade Masonic Lodge in Philadelphia. You have the Sweet Sixteen and Forty Eight lines of magnetism pictured together.

Ed Leedskalnin was a mason. Ed's symbol for magnetic current can be found of the door knocker from the Masonic lodge.

 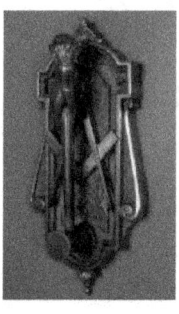 

Images of the mason lodge from
http://www.flickriver.com/photos/rkimberly/tags/masonic/

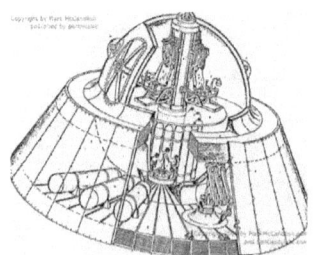

The image to the left if from a US made alien reproduction vehicle drawn by Mark Mc Candlish using an array of 48 capacitors in a circle to create the 48 lines of magnetism known by the Masons.

ED's Magnetic Current information from
http://www.leedskalnin.com/

Millions of people all over the world are being fooled by the non-existing electrons. Here is how the electrons came into existence. Thomson invented an imaginary baby and called it an electron. Rutherford adopted it and now the men with the long hair are nursing it.

The electron has a brother and its name is proton, but it is heavy and lazy. It remains stationary in the middle, but the electron has to run around it. To the electrical engineers the positive electricity is everything, the negative electricity is nothing, but to the physicists the negative electricity is everything, and the positive electricity is nothing. Looking from a neutral standpoint, they cancel each other, so we have no electricity, but we have something. If we do not know how to handle the thing that comes through a wire from a generator or a battery, we will get badly shocked. Read the booklet "Magnetic Current" then you will know what the thing is, and the way it runs through a wire.

The invention of an electron came by a tricky method in using electricity in a vacuum tube. Normally whether it is a generator or a battery, the positive terminal will have to be connected to the negative terminal, but in the vacuum tube two batteries with different strength were used, the smaller battery was connected normally, but the larger battery's negative terminal was connected to the smaller battery's negative terminal, and the positive terminal was left alone.

That connection gave the negative terminal a double dose of strength, and so it became hotter and could push more. It was called cathode and the positive terminal anode, and the electricity that passed from the cathode to the anode was called electrons.

In case the inventor had used normally direct methods to find out what the electricity was he would have found out that the positive and negative electricity is in equal strength, and are running positive electricity against the negative electricity.

That can be seen by connecting each of two pieces of soft iron wire with each terminal of a car battery and then by putting together and pulling away each loose end of the soft iron wire.

More sparks can be seen coming out of the positive terminal than from the negative terminal. This direct method is more reliable than the tricky method in the vacuum tube. The trouble with the physicists is they use indirect and ultra-indirect methods to come to their conclusions.

If the inventor of electrons had a vacuum tube in which his electrons could run close to the top of the vacuum tube from the west side of the cathode to the east side of the anode and then would hang a vertically hanging magnet that is made from three-inch long hard steel fishing wire, and then hang one magnet pole at one time right on top in the middle of his stream of electrons, then he would have seen the north pole magnet swinging north, and the south pole magnet swinging south. The same thing will happen if the magnets are held above any wire where the electricity is running through. Those two vertically hanging magnets prove that the electricity is composed of two different and equal forces.

Another way to prove this is to connect a flexible wire loop east end of the wire with positive battery's terminal, west end with negative terminal, raise the loop one inch above the floor. Put U shape magnet one inch from loop, North Pole south side of the loop. The North Pole magnet will pull in the loop. Put the South Pole magnet in the same place. It will push the loop away. Put the South Pole magnet north side of the loop, this time it will pull the loop in. Put the North Pole magnet in the same place, it will push the loop away. This indicates that electricity the same as a magnet bar is composed of two equal forces, and each force is running one against the other in whirling right hand twist, but those forces in the wire have higher speed, and both forces are coming out across from the same wire. One of the forces is North Pole magnets and the other is South Pole magnets. They are the cosmic forces. Your electric motor is turned around on its axis by north and South Pole magnets. Even you could not start your car without the north and South Pole magnets.

If electricity is made with north and South Pole magnets and the electric motor is turned around on its axis by the north and South Pole magnets, as is the fact, then this will bring up a question, where then are those Thomson electrons. They are not around the electric motor. The plain answer is they are non-existing.

Ed's secret. Overlay Ed's two books and unlock the key to magnetic current. North and South Pole magnetic current, + and - connector.

MAGNETIC CURRENT	Mineral, Vegetable and Animal Life	
		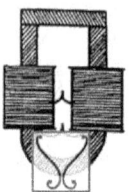
	Perpetual Motion Holder	
By EDWARD LEEDSKALNIN ROCK GATE HOMESTEAD, FLORIDA U S A	By EDWARD LEEDSKALNIN ROCK GATE HOMESTEAD, FLORIDA U. S. A.	

John Keely's Dynasphere
There are many different Music scales, the modern 440 Hz scale is not correct as it should 432 Hz.

DIATONIC SCALE NUMERICAL FREQUENCIES:

288	324	360	384	432	480	540	576
D	E	F	G	A	B	C	D

Note: "A" slightly adjusted

This is an example of harmonic relations between 22.5m, 24mm and 25.71428571mm in values when designing anything you may wish to invent. There are different combinations you can use but here is a start for you. With an outer sphere of 12 inches or 288mm in Diameter (24mm inch x 12=288 musical note D) and the inner sphere is 8 inches or 180mm in diameter (22.5mm inch x 8 = 180 musical note F).

The outer sphere and the inner sphere 288/180 = 1.6 which is Phi the Golden Ratio or Golden Mean as it goes under several names. The smallest sphere is 48mm in diameter and the Log of (24mm inch x2) 48 is 1.681.

The seven inner tubes inside Keely's device are 3 inches long each 25.71428571mm inch x 3 = 77.14285714 x 7 tubes = 540mm or musical note C. The 21 outer tubes are clearly smaller so 25.71428571mm inch x 2 = 51.42857143 x 21 tubes = 1080 musical note C one octave above the inner seven tubes.

Better pictures can be found on Dale Ponds website
http://www.svpvril.com/DynaspherePix/index.html

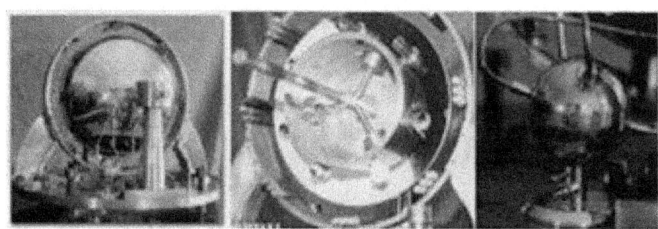

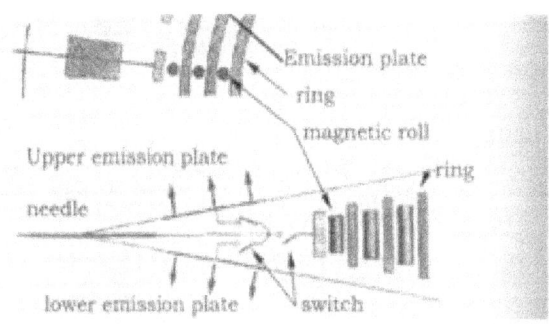

The picture to the left is a section of the Searl UFO You can see the magnetic rolls and in the lower section these three magnetic rolls are shown with varying height but, they are all the same length but on a different scale. As with the different musical scales throughout history, we also have a system of measurement that is a series of musical scales.

Sony corporation use 1080 HZ for their steady shot in there digital video cameras for perfect image resolution 25920 / 24 = 1080, for perfect resonance, the same principle applies to the 1080 HD High Definition televisions. Prior to 1080 Hz they used 720 HZ which is one complete 360 backwards because that's how they trick you into thinking you are getting a better model because of the picture quality when they already knew that the key the perfect resolution from day ONE of their manufacturing because of this science. All of this can easily be proved, with normal inbox specifications from the factory.

Most Electrical corporations know this like General Electric and Hertz. All the measuring system is the best-kept industrial secret but the consequences of it remaining secret is catastrophic to the people and the planet. Why should people have to pay for the price fixing and secret keeping when you will see the truth is before you?

Intelligent Designs best-known ratio of 1.618 origins in are found in the Fibonacci sequence, this also reveals many more ratios that are just as important. As the sequence, increases the ratios change ever so slightly this is because I believe entangled in the sequence each individual metric of the ninety measuring systems are entangled through the vortex.

People should ask "What creates the vortex, what makes it spin?" If you think of a blender you need some type of shape to create the movement and in this case it is the Swastika. This symbol belongs to different religions like Buddhism and Hinduism is the starting point of the vortex of information creating our reality.

The Fibonacci sequence. The Fibonacci system is another wonder of nature it id the order in which a rose grows it petals and the order that couples of rabbits are born from one couple in a year. It is extremely important in the development of anything to do with Vedic Physics. Every number of the sequence is the sum of the two previous numbers: for instance 1+1=2, 2+1=3, 3+2=5, and 5+3=8 without any possibility of reaching a final number.

1,1,2,3,5,8,13,21,34,55,89,144,233,377,610,987,1597,2584 etc.
The Fibonacci system and musical intervals.

1:1 Unison 5:3 Major Sixth
2:1 octave 8:5 minor sixth
3:2 Perfect 5th 13:8 Major Sixth

The ratios oscillate between major and minor sixth when approaching the harmonic sixth. Keely said "The rhythmic relations in which force acts are everywhere, under all conditions, and at all times, the same. They are found experimentally to be universally expressible by the mathematical relations of thirds. There are many calculations and applications that require the use of the Fibonacci system and without exception is the calculation of ratios for the Harmonic relationships of nature.

$3 \div 2 = 1.500000000$ $34 \div 21 = 1.619047619$ $377 \div 233 = 1.618025751$
$5 \div 3 = 1.666666667$ $55 \div 34 = 1.617647059$ $610 \div 377 = 1.618037135$
$8 \div 5 = 1.600000000$ $89 \div 55 = 1.618181818$ $987 \div 610 = 1.618032787$
$13 \div 8 = 1.625000000$ $144 \div 89 = 1.617977528$ $1597 \div 987 = 1.618034448$

Harmonic balance is achieved by finding the ratio that multiplies and divides evenly into the numbers related to sacred geometry.

Golf balls had 216 dimples on the circumference and now some have 432. Designs must be equal to nature to have a good day on the course.

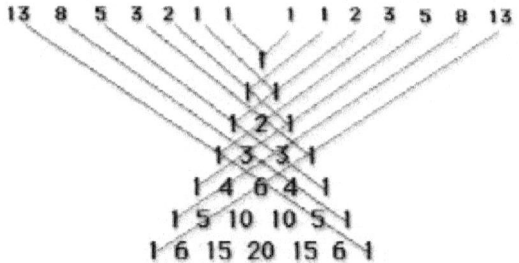

The Fibonacci Numbers in Pascal's Triangle

Pascal's Triangle was originally developed by the ancient Chinese, but Blaise Pascal was the first person to discover the importance of all of the patterns it contained the ratios of the Fibonacci sequence are known as the Golden Ratio which is known by many other names such as the golden mean, the divine proportion, or the golden proportion. It is represented by the Greek letter Phi (φ), and is an irrational mathematical constant approximately equal to 1.6180339887

No state of motion ever began or ever ended.

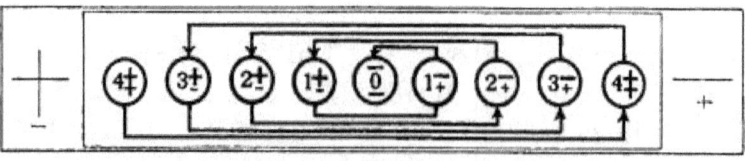

Tan 72·6 = Tan(30) Tan = 3·07 Tan θ   Tan 72·6 = Sin 72·6 / Tan θ   Tan 72·6 = Tan θ
Sin 72·6 = Sin(30) Sin = ·951 Sin 30 Sin θ   Sin = ·951 Sin = ·951   Sin = ·951 Sin 72·6 = Sin θ

1	2	3	4	5	6	7	8	9	10	11	12	13	14	
90	180	(270)	360	450	540	630	720	810	900	990	1080	1170	1260	0
36	72	108	144	180	216	252	288	324	360				28	

1476 1512 1548 1584 1620 1656 1692 1728 1764 1800
1116 1152 1188 1224 1260 1296 1332 1368 1404 1440
756 792 828 864 900 936 972 1008 1044 1080
396 432 468 504 540 576 612 648 684 720
36 72 108 144 180 216 252 288 324 360

Tan t = 3·07703537   Tan t = 0·706542528
Sin t = 0·910565616   Sin t = 0·587763252

Sin=Tan=θ ⑤

and...
one mm...

Nine + Zero Are the Same
Eg) $18 - 1 = 9$ Eg) $21 \div 3 = 3$
$1 \cdot 9 \cdot 2 = 1 \cdot 1 \cdot 2 = 9$ $1 + 8 = 9$

$\frac{2}{9}$ $\frac{3}{9}$
$\frac{3 \cdot 9}{9}$
$45, 36, 27$

All Add To 1
$64 = 6 + 4 = 10 \to 1 + 0 = 1$
$55 = 5 + 5 = 10 \to 1$
$46 = 4 + 6 = 10 \to 1$
$3 \times 37 = 111 \to 1 + 1 + 1 = 3$
$28 = 2 + 8 = 10 \to 1$
$19 = 1 + 9 = 10 \to 1$

All Add To 2
$56 = 5 + 6 = 11 = 1 + 1 = 2$
$47 = 4 + 7 = 11 = 2$
$38 = 3 + 8 = 11 = 2$
$29 = 2 + 9 = 11 = 2$

$3, 6, 9$
$3, 8, 8$
$3 \times 3 = 9$
$5 \times 6 = 30$
$13 \times 3 = 39$

7 7 7
7 7 7
$7 \times 5 = 35$ $7 \times 4 = 28$ $25, 16, 7$

The Pendulum
Swings Between
$1 + 9$

$6 = 15$ $5 = 14$
$6 = 24$ $5 = 23$
$6 = 33$ $5 = 32$
$6 = 42$ $5 = 41$
         $5 = 50$
         $5 = 59$

$4 = 13$
$4 = 22$
$4 = 31$
$4 = 40$

$3 = 12$
$3 = 21$
$3 = 30$

$2 = 11$
$2 = 20$

All Add To 3
$12 = 1 + 2 = 3$
$48 = 4 + 8 = 12 = 1 + 2 = 3$
$57 = 5 + 7 = 12 = 3$
$39 = 3 + 9 = 12 = 3$

All Add To 4
$5 \cdot 8 = 13 = 1 + 3 = 4$
$4 + 9 = 13 = 1 + 3 = 4$
$14 = 0 + 1 = 4 \ldots$

$3 + 9 = 12 = 1 + 2 = 3$
$18 - 9 = 9$
$3 \div 9 = 0.3$
$18 \div 9 = 2$

$0 / 6 = 0$

No Number Adds up to more than 9

The circle of nine where there is no number greater than nine.

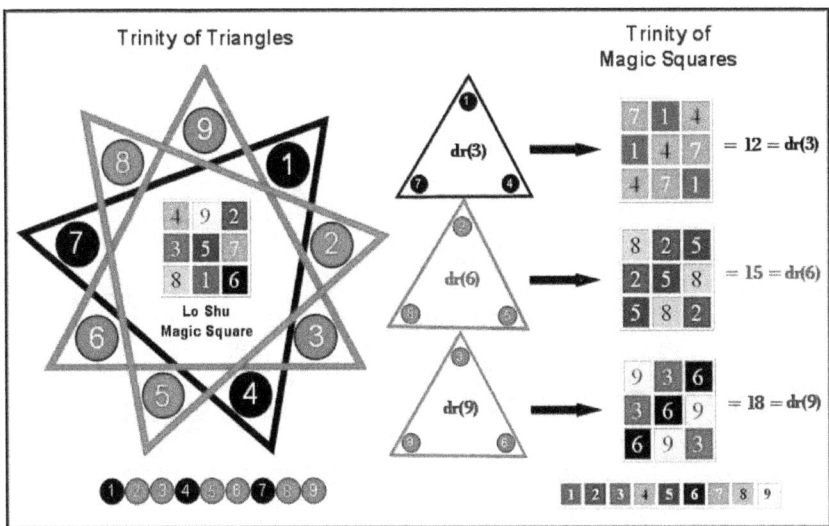

This is the same as the previous page but showing it using magic squares and the 3, 6 and 9 secret of the universe. Above image from www.primesdemystified.com

Intelligent Design

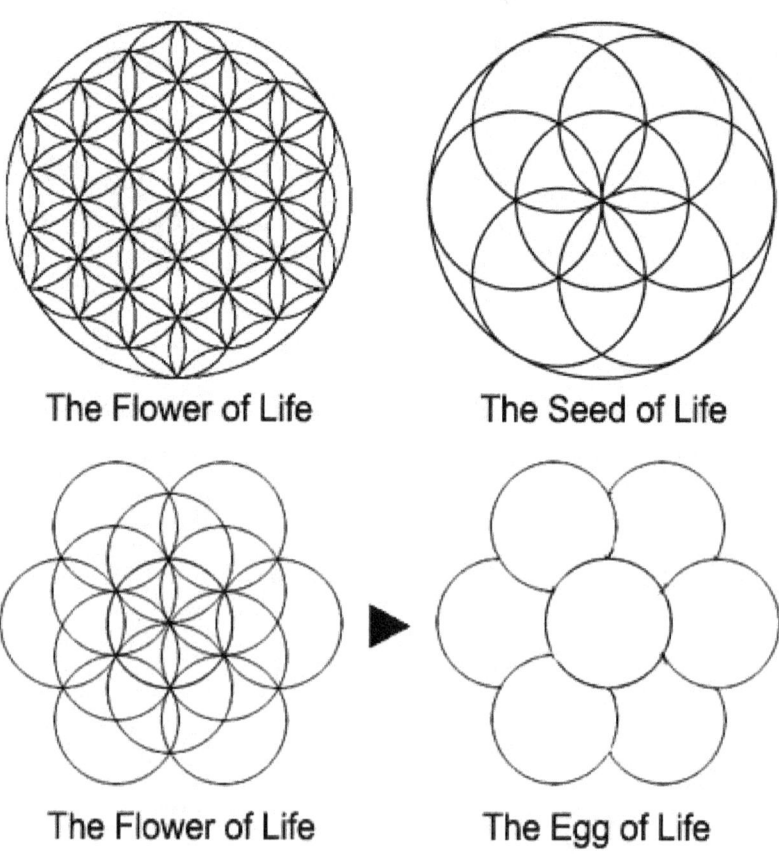

The Flower of Life

The Seed of Life

The Flower of Life

The Egg of Life

The flower of life Geometry contains all the Platonic solids and was used by Leonardo Di Vinci and many before.

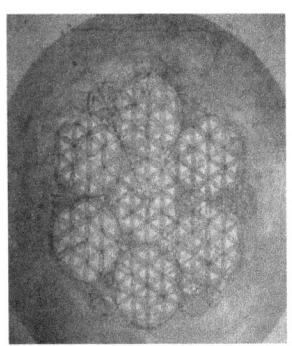 

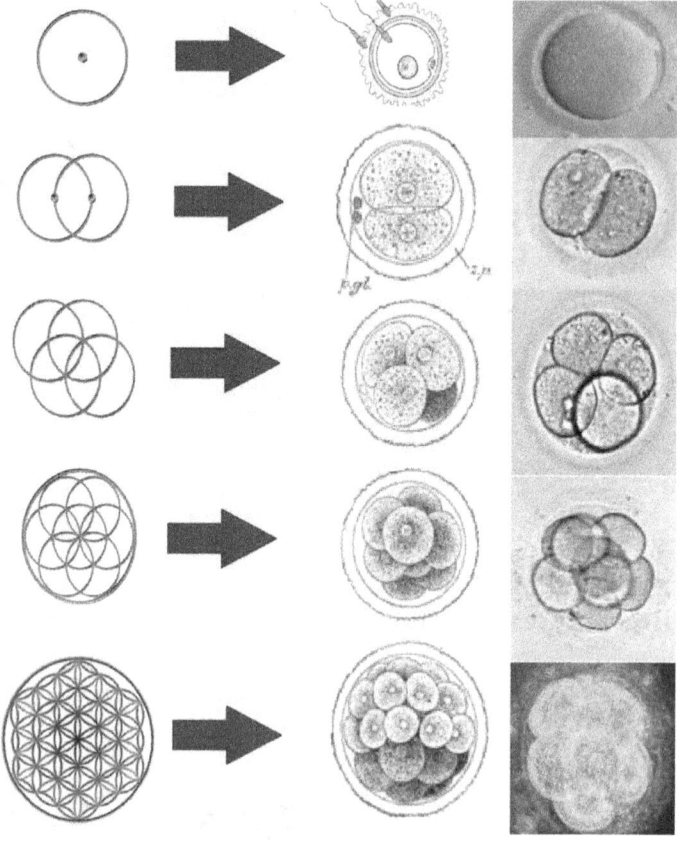

This symbol has been found in over 100 countries all over the world.

You can see how the flower of life is contained in all living cells.

http://transmissionsmedia.com/flower-of-life-and-cell-division-geometry-of-creation/

I have to say I truly admire the work of Robert Thomas and there should be more recognition of it. "The WORD made Manifest through Sacred Geometry" is truly a life-changing book and everyone should read it.

Genesis Model -Creation of the Seed of Life from the Star the Story of Creation.

Genesis 1:1 "In the beginning God created the heaven and the earth."

Comment: The Hebrew word for created means "a making out of nothing." "Nothing" can be thought of as a point, which fixes position and has no area. God is the infinite point, the centre of creation. This point relates to Hermetic Principle I, THE PRINCIPLE OF MENTALISM. The act of creating involves God mentally projecting equally in all six directions on the X-Y-Z axis at a particular radius, creating an inside and outside. This can be shown as a sphere. To keep the drawing in its simplest form, a two-dimensional representation is shown as a circle.

See Figure 3-1
Genesis 1:2 "And the earth was without form and void; and darkness was upon the face of the deep. And the Spirit of God moved upon the face of the waters."

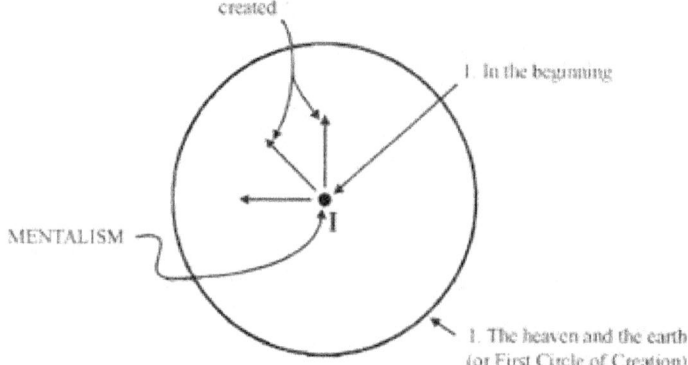

Figure 3-1. Genesis 1:1 (the numbers represent the verses)

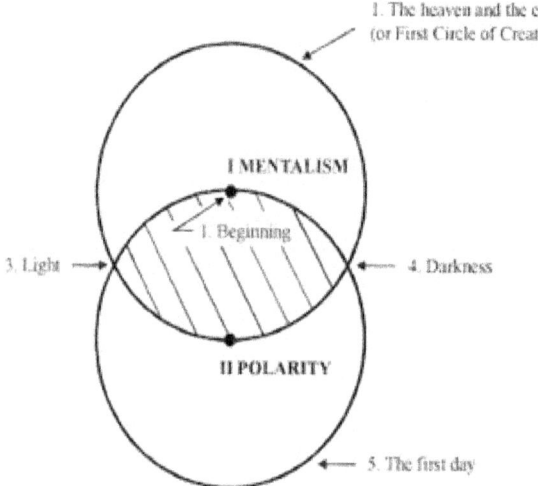

Figure 3-2. Genesis 1:3-5, "The First Day"

Comment: At this point the earth is formless, for God hasn't manifested on the circle of creation. God's Spirit must be present in order for substance and life to occur. Genesis 1:3-5, And God said, Let there be light: and there was light. 4 And God saw the light, that it was good: and God divided the light from the darkness. 5 And God called the light Day, and the darkness he called Night. And the evening and the morning were the first day."

Comment: "The first day" is represented by the first circle whose centre lies on the circumference of the Circle of Creation. See Figure 3-2. The centre of the circle relates to Hermetic Principle II, THE PRINCIPLE OF POLARITY. The two circles have the same radius. The circumference of "the first day" must intersect with point I which gives life to II. The circumferences of the circles intersect in two places, points 3 and 4. These points represent the two opposing aspects of I and II. Points 3 and 4 are the same distance away from I and II, both horizontally and vertically.

Point 3 is represented as light and 4 as darkness to match the verses. The common area of I and II form what is called a "vesica pisces" (shaded area).

Genesis 1:6-8 And God said, Let there be a firmament in the midst of the waters, and let it divide the waters from the waters. 7 And God made the firmament, and divided the waters which were under the firmament from the waters which were above the firmament: and it was so. 8 And God called the firmament Heaven. And the evening and the morning were the second day."

Comment: This passage is referring to the different levels of consciousness or planes of existence. The levels are divided so that only those of the proper vibration can reach the next level. An increasing rate of vibration is what divides the lower waters from the higher waters. The different levels are all called waters, so there are similarities between them. Verses 6 through 8 are referring to Hermetic Principle III, THE PRINCIPLE OF CORRESPONDENCE. As above, so below, as within, so without. Since above and below (or under) correspond with each other and are both described in verse 7, the intersection points of I and III are both listed as verse 7.

See Figure 3-3.

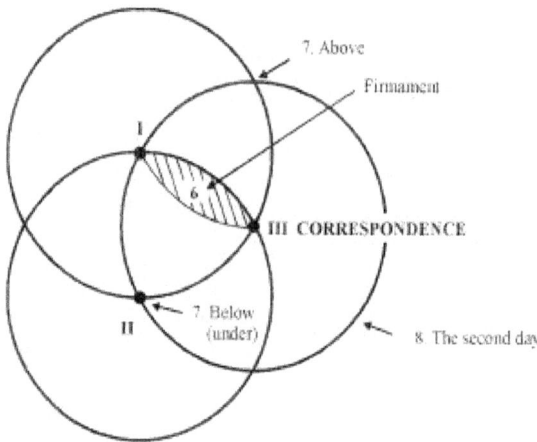

Figure 3-3. Genesis 1:6–8, "The Second Day"

Genesis 1:9-13 And God said, Let the waters under the heaven be gathered together unto one place, and let the dry land appear: and it was so. 1 And God called the dry land Earth; and the gathering together of the waters called the Seas: and God saw that it was good. 11 And God said, let the earth bring forth grass, the herb yielding seed, and the fruit tree yielding fruit after his kind, whose seed is in itself, upon the earth: and it was so. 12 And the earth brought forth grass and herb yielding seed after his kind, and the tree yielding fruit, whose seed was in itself, after his kind: and God saw that it was good. 13 And the evening and the morning were the third day."

Comment: This pertains to Hermetic Principle IV, THE PRINCIPLE OF CAUSE AND EFFECT. The mountains are forming and the topography of the earth is being shaped by natural processes. This is the cause. The effect is that the water will find its proper level.

In order for vegetation to multiply, it must bear fruit containing its seed. The seed-bearing vegetation created by God is the cause. The effect is the widespread growth of the plants. See Figure 3-4.

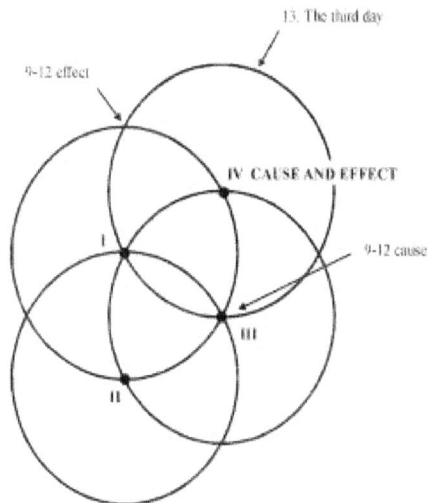

Figure 3-4. Genesis 1:9–13. "The Third Day"

Genesis 1:14-19 And God said, Let there be lights in the firmament of the heaven to divide the day from the night; and let them be for signs, and for seasons, and for days, and years: 15 And let them be for lights in the firmament of the heaven to give light upon the earth: and it was so. 16 And God made two great lights: the greater light to rule the day, and the lesser light to rule the night: *he made* the stars also. 17 And God set them in the firmament of the heaven to give light upon the earth, 18 and to rule over the day and over the night, and to divide the light from the darkness: and God saw that *it was* good. 19 And the evening and the morning were the fourth day."

Comment: Verse 14 states; let the lights serve as signs for the fixing of seasons, days and years. This is referring to Hermetic Principle V, THE PRINCIPLE OF RHYTHM, which controls the timing of the seasons and other rhythmic motion. This compares to the back-and-forth motion of the swinging pendulum. These two words are shown as the opposing aspects of rhythm.

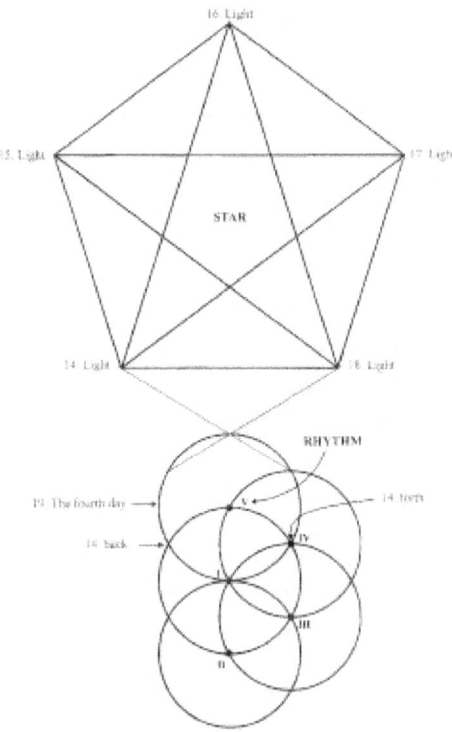

Figure 3-5. Genesis 1:14-19, "The Fourth Day"

See Figure 3-5. The main emphasis in verses 14 through 18 is lights. It is repeated at least once in all five verses. This relates to the five points of light, the five-pointed star, or the pentagram. God created the sun and the other lights in the heavens to nourish the earth, just as the pentagram energizes the Seven Days of Creation, "The Seed of Life."

Genesis 1:20-23 And God said, Let the waters bring forth abundantly the moving creature that hath life, and the fowl that may fly above the earth in the open firmament of heaven. 21 And God created great whales, and every living creature that moveth, which the water brought forth abundantly, after their kind, and every winged fowl after his kind: and God saw that it was good. 22 And God blessed them, saying, Be fruitful, and multiply, and fill the waters in the seas, and let fowl multiply in the earth. 23 And the evening and the morning were the fifth day."

Comment: The first three verses of the fifth day each repeat God's creation of all of life within the waters and in the air. The various creatures that live within these areas vary widely in physical and mental abilities, the whale being the most evolved. The brain size of the whale and dolphin are much greater than other animals. The more advanced the creature, the higher its vibratory rate. This relates to Hermetic Principle VI, THE PRINCIPLE OF VIBRATION. Water and air are two of the four basic elements of earth, air, fire, and water. Water is a contractive force, whereas air is expansive. They are shown as the opposing aspects of vibration in Figure 3-6.

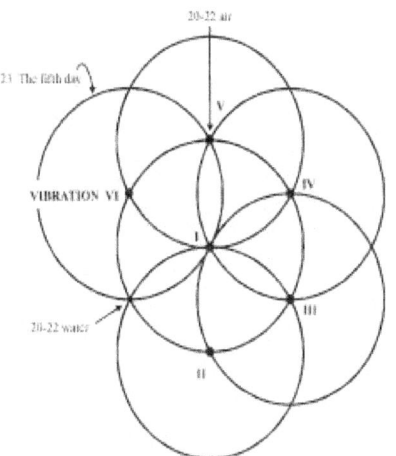

Figure 3-6. Genesis 1:20-23, "The Fifth Day"

Genesis 1:24-31 And God said, Let the earth bring forth the living creature after his kind, cattle, and creeping thing, and beast of the earth after his kind: and it was so.

25 And God made the beast of the earth after his kind and cattle after their kind, and everything that creepeth upon the earth after his kind: and God saw that it was good.
26 And God said; Let us make man in our image, after our likeness: and let them have dominion over the fish of the sea, and over the fowl of the air, and over the cattle, and over all the earth, and over every creeping thing that creepeth upon the earth. 27 So God created man in his own image, in the image of God created he him; male and female created he them.

28 And God blessed them, and God said unto them, Be fruitful, and multiply, and replenish the earth, and subdue it: and have dominion over the fish of the sea, and over the fowl of the air, and over every living thing that moveth upon the earth?
9 And God said, Behold I have given you every herb bearing seed, which is upon the face of all the earth, and every tree, in which is the fruit of a tree yielding seed; to you it shall be for meat. 30 And to every beast of the earth, and to every fowl of the air, and to everything that creepeth upon the earth, wherein there is life, I have given every green herb for meat: and it was so. 3] And God saw everything that he had made, and, behold, it was very good. And the evening and the morning were the sixth day."

Comment: Verses 26 and 27 refer to God creating man in His own image. Male and female, He created them. According to ancient philosophy, "Man is the measure of all things," and all things have their masculine and feminine properties. This corresponds to Hermetic Principle VII, THE PRINCIPLE OF GENDER. See Figure 3-7.

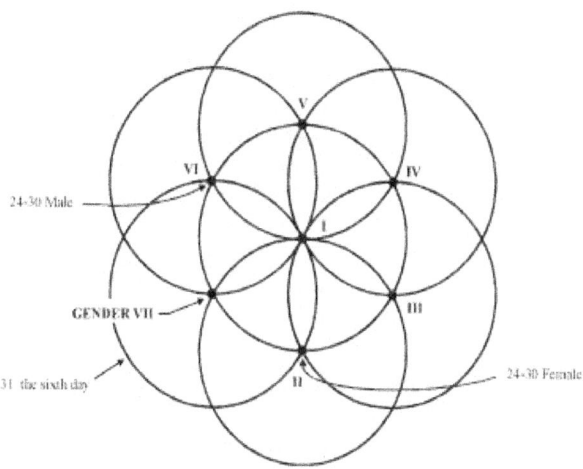

Figure 3-7. Genesis 1:24-31, "The Sixth Day"

Genesis Chapter 1 has described the design of the Seed of Life from which all of life is derived, and the pentagram (star) which nourishes it. The pattern is now complete and ready to expand -the heavens and earth were finished with this stage. Figure 3-8 shows the complete design with all 31 labelled verses. A second circle is shown which encompasses all six days of creation and is noted as the Second Circle of Creation. Its radius is twice that of the first. This entire book is based on the Genesis Model and the growth process of the Seed and Star. All of life contains this seed.

Forces within the Seed of Life

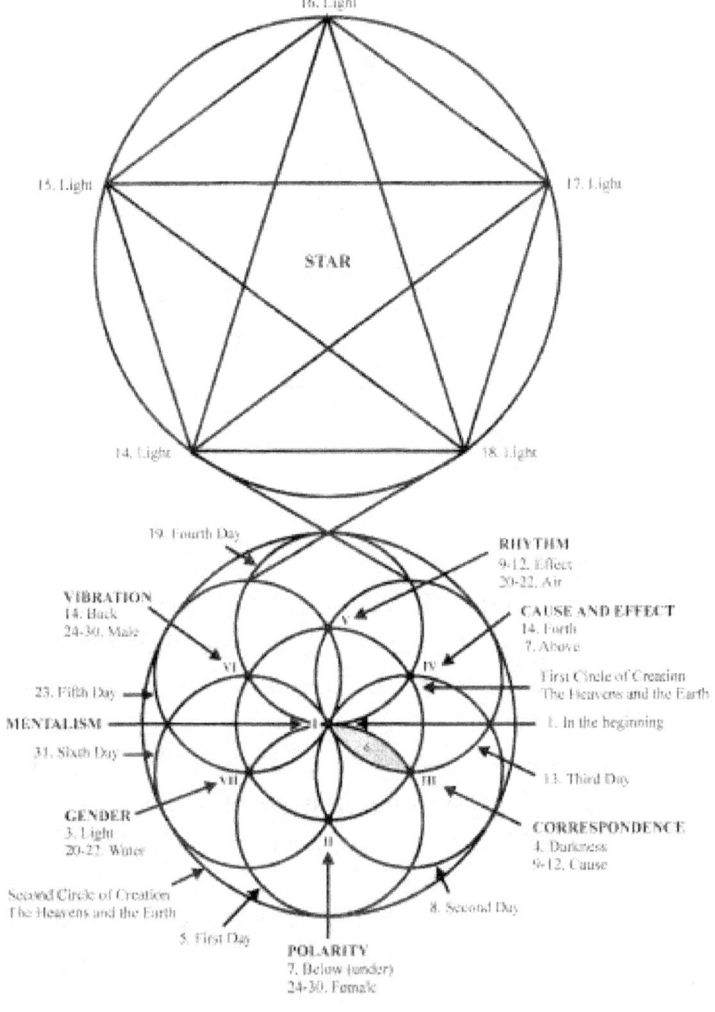

Figure 3-8. Genesis 1:1–31, "The Genesis Model"

Refer to Figures 3-8 and 3-9 to see the relationship between the Hermetic Principles. As stated previously, when comparing I and II (Figure 3-8), point VII (3. light) is the opposing aspect of point III (4. darkness). The line between VII and III is in repulsion. This is shown by the arrows pointing away from each other.

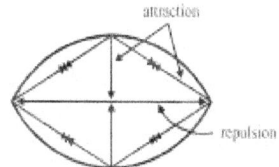

Figure 3-10. Forces within the Vesica Pisces

See Figure 3-9. Forces in repulsion have a like charge and, in this case, both points are negative. When comparing VII and III, with these points being the centre of the two circles, their circumferences intersect at two points, I and II. They are in common or in attraction with VII and III. Point II is also point 24 to 30 (female) in common with VII, Gender, and II (7. Below) is in common with III, Correspondence. Point I is in attraction with all the Hermetic Principles. Attraction is designated as arrows pointing toward each other. Forces in attraction have unlike charge, one being positive and the other negative.

All of the Hermetic Principles have this relationship. Figure 3-9 shows the forces involved. Hermetic Principle 1 is a black hole or implosion well and attracts all of the vertexes of the hexagram. It is also a white hole which sends out the divine spark at certain intervals to energize the seed. This energy originates from the star.

The black hole attracts both the positively and negatively charged particles and are compressed into etheric matter and expelled during the white-hole stage. The feminine aspects of the seed are the circles. The straight lines forming the hexagram are its masculine counterparts.

Star of David Figure 3-9 shows that every adjacent Hermetic Principle forming a hexagon is in attraction. This would require one being positive and the other being negative. Every other Hermetic Principle forming two opposite triangles is called a hexagram or Star of David is in repulsion, all points being negative or positive. The triangle is the simplest enclosed two-dimensional form and is structurally stable. The triangle resists all movement in any direction. Every third Hermetic Principle at the opposite side is in attraction. One point is negative and the other positive. The attraction and repulsion forces are in a constant battle for dominance, each tending to balance the other. The resulting stress allows the Seed of Life to draw energy from the ethers to form matter in an orderly interference pattern using Sacred Geometry.

Hence, the "Seven Days of Creation." God expresses on the Circles of Creation through the six Hermetic Principles forming the Star of David. Each point of the two triangles represents one of the three aspects of God forming the trinity.

Path of the Flaming Sword as Electrical Circuit

Genesis 3:22-24: 22 And the Lord God said, Behold, the man is become as one of us, to know good and evil: and now, lest he put forth his hand, and take also of the tree of life, and eat, and live forever: 23 Therefore the Lord God sent him forth from the garden of Eden, to till the ground 24 from whence he was taken.

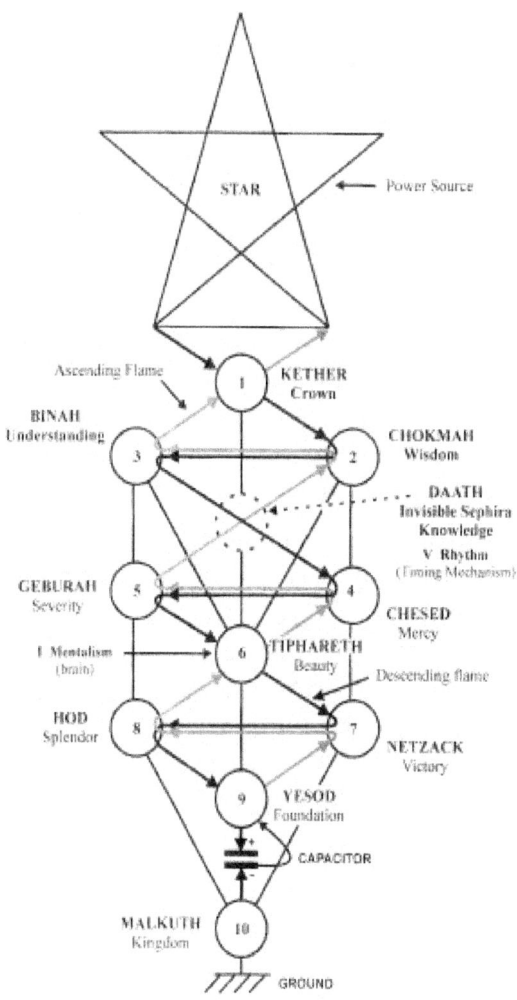

Figure 4-5. The Path of the Flaming Sword

So he drove out the man; and he placed at the east of the garden of Eden Cherubims, and a flaming sword which turned every way, to keep the way of the tree of life." The fall in consciousness opened their eyes to duality experiences, and to know good and evil, and to suffer the consequences. Mankind lost the knowledge of the complete path of the flaming sword which turned every way, to keep the way of the Tree of Life. Books on the Qabalah show only one-half of the cycle of the path of the flaming sword descending through Kether to Malkuth, in a crossing pattern. This is an electrical circuit which requires a positive and negative or neutral line.

There must be a return path to complete the circuit to fulfil the requirement the path of the flaming sword turns every way, which means both up and down. Figure 4-5 shows the circuit. Replace the Ain, Ain Soph, and Ain Soph Aur with the pentagram or star. Daath or its other name, V Rhythm, controls the timing of the spark from the star. V is the first Hermetic Principle the flaming sword encounters on its downward path. Yesod or II Polarity is the last.

The following text is from "The Dimensions of Paradise" By John Michell

The ancient Greeks were well aware of [torsion] energy, calling it "aether" and understanding that it is directly responsible for universal manifestation. In the 1950s Russian scientist Nicolai Kozyrev conclusively proved the existence of this life-giving subspace energy, demonstrating that, like time, it flows in a sacred geometric spiral resembling the involutions of a conch shell that has been called phi, the Golden Mean, and the Fibonacci sequence.

Several of the incidents and parables in the New Testament story of Jesus are known to have been adapted from earlier writings, and some of these have hidden meanings that the Christian Gnostics interpreted using the same cabalistic methods that the Jews apply to the exegesis of their own scriptures. Most obviously numerical is the tale of the miraculous catch of 153 fishes, which occurs in the last chapter of St. John's Gospel. Why there should have been exactly 153 fishes in the net that the Apostles cast into the Sea of Tiberias is a question that has puzzled commentators from early Christian times. A clue that previous writers have noticed is that two of the key words in the story, fishes, and the net, each have the value by gematria of 1224, and 1224 is 8 times 153.

Following up this clue we are led on to reconstruct the figure of sacred geometry that must originally have accompanied the story of the 153 fishes. It develops in three stages, reflecting the order of events in John 21.

After the Crucifixion, Simon Peter went fishing from a boat in the Sea of Tiberias, taking with him six of the other disciples. They fished all night but caught nothing. In the morning they saw the risen Jesus on the shore but failed to recognize him. He called out that they should cast their net on the right side of the boat. Having done so, they were unable to draw it out for the multitude of fishes in it.

John then recognized Jesus and told Simon Peter, who put on his fisher's coat and jumped into the sea. The other disciples followed him in the boat to the shore, which was about 200 cubits away, dragging the net with the fishes. When Simon Peter drew it to land it was found to be "full of great fishes, an hundred and fifty and three: and for all there were so many, yet was not the net broken."

The number of Simon Peter, is 1925, so Peter can be represented by a circle with circumference 1925 and diameter 612½ or 612. This is appropriate because 612 is the number of the Good Shepherd, and that is the title which Simon Peter inherits when, following the incident of the 153 fishes, he is told three times by Jesus, "Feed my sheep."

Figure 67. Seven circles, representing six disciples with Simon Peter at their center, pack into the circular boat of diameter 1224.

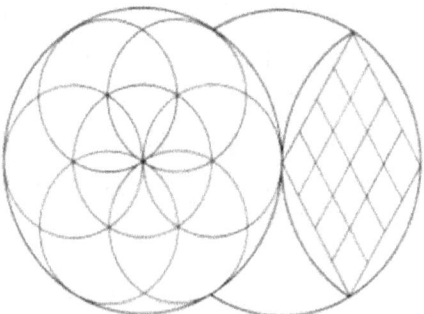

Figure 68. A net formed by a rhombus within a Vesica Piscis is cast on the right side of the boat, catching 153 fishes.

Six more circles of the same dimensions are drawn for the six other disciples, and the seven are packed together in the most economical way (figure 67) and placed inside the circular boat, like the coracle of the Celtic saints, the diameter of which is 1224.

The disciples are told to cast their net on the right side of the boat. This is done geometrically by placing the compass point on the circumference of the circular boat and drawing an arc of another circle with the same radius, containing a Vesica Piscis (vessel of the fish). The rhombus within it is divided up into sixteen smaller diamond shapes (figure 68). Its width being 612, each of its sixteen divisions has a width of 153. They represent sixteen small fishes making up a greater seventeenth, and here again the number 153 is brought out, for 153 is the sum of the numbers from 1 to 17.

The measure around the four sides of the greater rhombus-fish is 2448 or 1224 + 1224 or the net, plus fishes. Thus the net full of 153 fishes is illustrated in number and geometry. It is a traditional practice among teachers of esoteric philosophy to set forth their doctrines in the guise of simple parables, which amuse children, enrich popular mythology, and, for those who understand the science of interpreting them, illustrate various cosmological processes. The themes that are adopted by hagiographers and composers of sacred legends are those that occur spontaneously in different times and cultures and can therefore be called archetypal.

Thus the founders of Christianity took certain episodes in universal folklore and made Jesus their central figure. In the tale of the 153 fishes he plays the part of the shamanic man of miracles whose traditional functions include bringing good luck to hunters or fishermen. By interpolation of names and numbers this story was made to reflect the construction of a geometric diagram with cosmological significance, by reference to which the gnostic masters were able to demonstrate to initiates the basic truth behind the Christian legend.

John Mitchell's description is the correct interpretation of this bible passage.

The Lucifer experiment
The Devil is in the detail

The Lucifer Unraveling
Two centers, instead of one

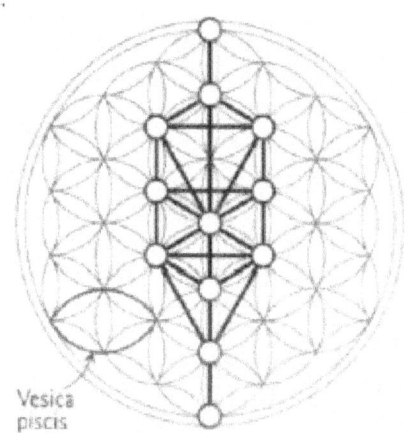

Fig. 2-24. Tree of Life with two extra circles.

The concept of Lucifer is an experiment for humans to live with free will. How could free will exist unless consciousness created this possibility? All creations are Gods creations, Lucifer did not create free will but it was because of his actions that free will now exists. In the study of sacred geometry nothing was created without reason and Lucifer is an experiment in the duality of consciousness (separation of the ONE into TWO.) The Freemason worship Lucifer it is their god of love and light, the god of magnetism.

Duality is not oneness or in Krishna Consciousness. Do not be trapped by their thinking.

"That which we must say to the **CROWD** is: we worship a god, but it is the god that one adores without superstition. To **YOU** Sovereign Grand Inspectors General, we say this, that you may repeat it to the brethren of the 32nd, 31st and 30th degrees – the **MASONIC RELIGION** should be, by all of us initiates of the *high* degrees, maintained in the purity of the **LUCIFERIAN** doctrine. If Lucifer were not god, would Adonay (Jesus)... calumniate (spread false and harmful statements about) him?... YES, **LUCIFER IS GOD...**"

A.C. De La Rive, La Femme et l'Enfant dans la Franc-Maçonnerie Universelle (page 588).

Black and white floor

General Albert Pike

"...Yes, lucifer is god, and unfortunately Adonay is also God, for the eternal law is that there is no light without shade, no beauty without ugliness, no **white without black**, for the absolute can only exist as **two gods**, darkness being necessary for light to serve as its foil, as the pedestal is necessary to the statue, and the brake to the locomotive..."

- General Albert Pike, Grand Commander, sovereign pontiff of universal freemasonry, giving instructions to the 23 supreme councils of the world. Recorded by A.C. De La Rive, La Femme et L'enfant dans la Franc-Maçounerie Universelle, Page 588. Cited from: The question of freemasonry. (2nd edition 1986 by Edward Decker pp12-14.)

XXXVI.

The light we perceive is a part only of the infinite light, the few solar rays which correspond with our visual apparatus. The sun itself is a lamp adjusted to our dim sight ; it is a luminous point in that space which would be darkness to the eyes of our body, but is resplendent for the intuition of our souls.

XXXVII.

The word magnetism expresses the action and not the nature of the great universal agent which serves as mediator between thought and life. This agent is the infinite light, or, seeing that the light is itself only phenomenal, it is rather the light-bearer, the great Lucifer of Nature, the mediator between matter and spirit, the first creature of God, but termed the devil by impostors and the ignorant.

XXXVIII.

What is more absurd and more impious than to attribute the name of Lucifer to the devil, that is, to personified evil. The intellectual Lucifer is the spirit of intelligence and love ; it is the Paraclete, it is the Holy Spirit, while the physical Lucifer is the great agent of universal magnetism.

XXXIX.

To personify evil and exalt it into an intelligence which is the rival of God, into a being which can understand but love no more-this is a monstrous fiction. To believe that God permits this evil intelligence to deceive and destroy his feeble creatures is to make God more wicked than the devil. By depriving the devil of the possibility of love and repentance, God forces him to do evil. Moreover a spirit of error and falsehood can only be a folly which thinks, nor does it deserve indeed the name of spirit. The devil is God's antithesis, and if we define God as He who is we must define His opposite as he who is not.

XL

We must seek for the spirit of Dogmas, while receiving their letter in its integrity as the priestly sphinx transmits it.

Freemasons are apparently just a society of secrets not a secret society but this is their geometry and beliefs. The name Lucifer comes from the Latin lucem ferre or "light bearer." The poetic name given to the morning star (Venus) was "Lucifer," from the Latin "lucem ferre," the bringer or bearer of light, which alludes to a Greek god named, Phosphorus.

LUCIFER
Theosophical magazine,
DESIGNED TO "BRING TO LIGHT THE HIDDEN THINGS OF DARKNESS."
EDITED BY H. P. BLAVATSKY AND MABEL COLLINS.
Lucifer volume 1 through volume 10 September 1887 August 1892

The Lucifer magazines covering the above dates contain some 4000 pages of information for anyone who is really interested in knowing.

"Lucifer, the *Light-bearer* ! Strange and mysterious name to give to the Spirit of darkness! Lucifer, the Son of the Morning! It is he who bears the Light, and with its splendors intolerable, blinds feeble, sensual or selfish souls? Doubt it not!"

Albert Pike, Morals and Dogma of the ancient and Accepted Scottish Rite of Freemasonry, p321, 19th Degree of Grand Pontiff

Remember back to Ed Leedskalnin and the Coral Castle, and then you will know this is true.

"When the Mason learns that the key to the warrior on the block is the proper application of the dynamo of living power, he has learned the mystery of his craft. The seething energies of Lucifer are in his hands and before he may step onwards and upwards he must prove his ability to properly apply (this) energy."

'Lost Keys of Freemasonry' page 48, Manley P Hall 33rd degree.

One of the most hidden secrets involves the so-called fall of the Angles. Satan and his rebellious host will thus prove to have become the direct Saviours and Creators of define man. Thus Satan, once he ceases to be viewed in the superstitious spirit of the church, grows into the grandiose image. It is Satan who is the god of our planet and the only God. Satan (or Lucifer) represents the Centrifugal Energy of the Universe, this ever-living symbol of self-sacrifice for the intellectual independence of humanity.

Sister H.P Balavatsky
The secret Doctrine.

The Human Machine
Master numbers 11, 22, 33
Found In Human DNA

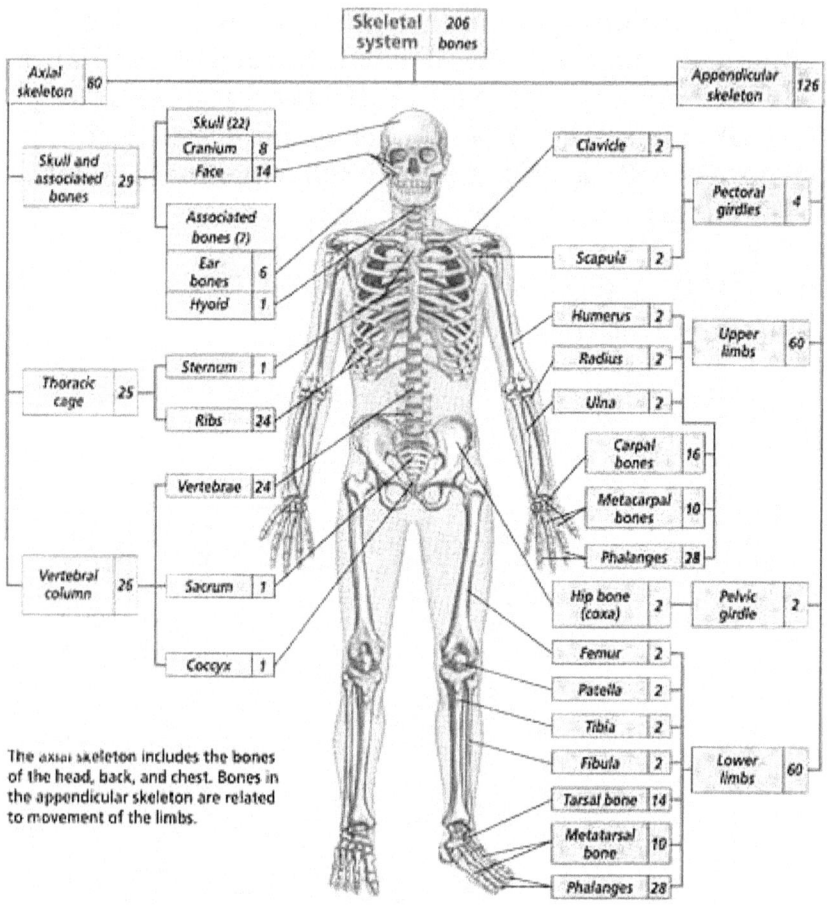

You are a machine - a complex orchestration of three-dimensional fractal patterns, "created primarily for mundane slave labour" and impregnation of your female counterpart - to perpetuate future generations of your worker species.

All senses in the body are routed to the brain via the nervous system. When signals, sent by nerve endings, reach the brain, neurons within it fire in three-dimensional fractal patterns, processing the information. For every memory, there is a fractal pattern which represents it that is stored within your brain. Sounds simple enough?

The brain is a three-dimensional fractal interface for the soul.

Your skeletal system is comprised of many bones. In particular, your skull has 22 bones and your back has 33 vertebrae. Your rib cage has 11 true ribs on each side. There is a twelfth rib, on each side, but it is not considered a true rib, because it is not bone. 11, 22, and 33 are precise or 'master' numbers. They are also multiples of 11. I couldn't even begin to calculate the odds of all of these numbers being *encoded within your DNA* through any other process other than "intelligent thought".

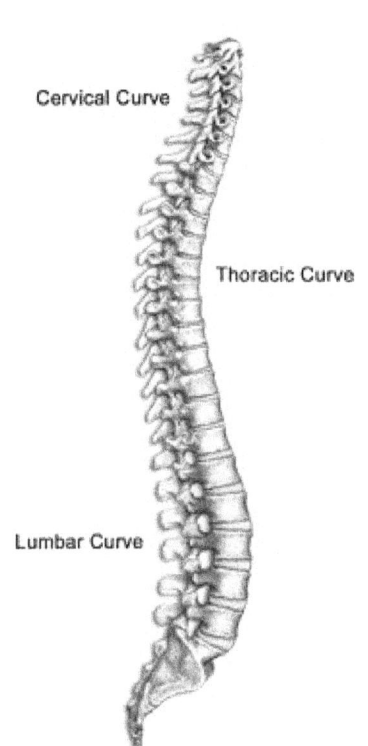

Master Numbers within Your DNA The Vertebral Column

The human body is an amazing thing - or more accurately stated - *machine*. It is fully functional and extremely versatile. It offers a method of locomotion to that which is not of the physical (i.e. the spirit or soul). There are several sub-systems which make up the physical body.
The major systems are the skeletal, muscular, nervous and cardiovascular. The skeletal system provides the foundation and all of the other systems are combined with it to make a biological machine.

The nervous system provides the sensation of touch, taste, smell, sight and much more, all of which are needed by this wondrous machine to interact with the physical universe. Your back is made up of a series of bones called 'vertebrae'. Together they form a flexible column. There are a total of thirty-three vertebrae and they are grouped under the names cervical, thoracic, lumbar, sacral, and coccygeal. There are seven in the cervical region, twelve in the thoracic, five in the lumbar, five in the sacral, and four in the coccygeal.

Freemasonry should pop into your mind at this point. The number 33 is the highest degree within Masonry. As you look at the vertebrae in the column, the ones in the neck are smaller than the ones lower down. The reason for this is to provide a greater range of motion and flexibility for your skull.

The Skull

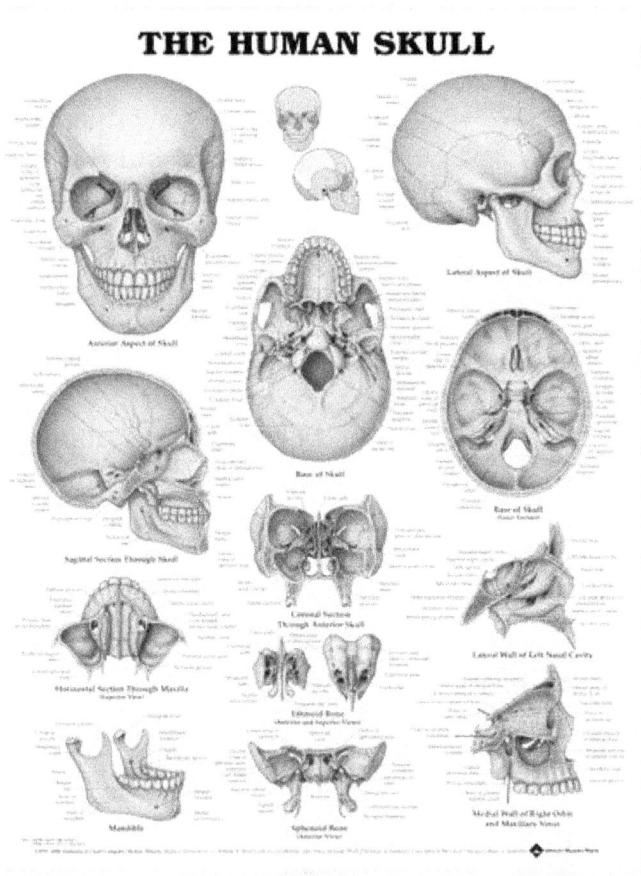

At the summit of the vertebral column is the skull. The skull is oval shaped. It is wider behind than in front. The bones it is comprised of are flattened or irregular. The only exception is the mandible (commonly known as the jawbone). These bones are immovably jointed together. The cranium lodges and protects the brain and consists of eight bones, while the skeleton of the face consists of fourteen bones:

The Ribs

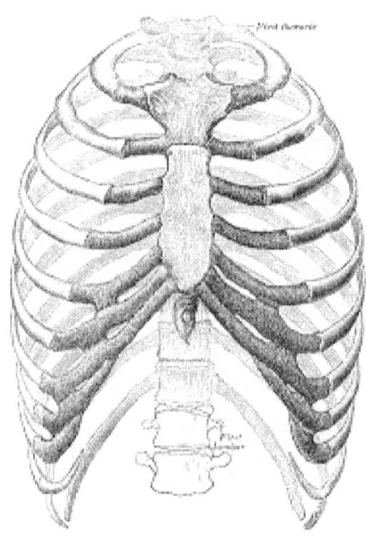

The ribs are comprised of elastic arches of bone. They form a large part of the thoracic skeleton. On either side, there are twelve, but this number may be increased by the development of a cervical or lumbar rib, or may be diminished to eleven.

Behind the vertebral column is where the first seven ribs connect. In front, they connect through the intervention of the costal cartilages, with the sternum. They are called true or vertebro-sternal ribs. The other five ribs are "false" ribs. The first three ribs have their cartilages attached to the cartilage of the rib above (vertebro-chondral). The last two are free at their anterior extremities and are referred to as floating, or vertebral, ribs.

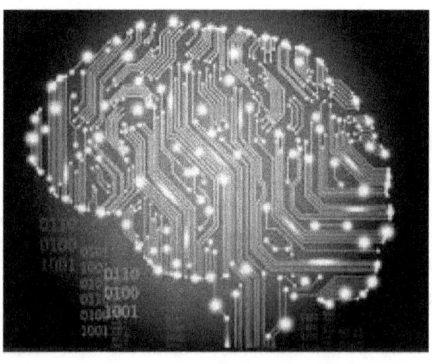

The Brain and DNA (Deoxyribonucleic Acid)

The human brain is most interesting. When any 'memory' is accessed, neurons within it fire in a *three-dimensional fractal pattern*. Examples of a three-dimensional fractal pattern are the outline of a cloud, a bolt of lightning, or the veins in your arms. Fractal Geometry, like everything else in this universe (and others) is based on mathematics, a medium known to be predetermined.

Everything about our physical bodies is described within our DNA. The gene is a unit of heredity. It is time for people to approach the human body from a spiritual perspective. If they would, they would begin to understand more about how it truly works, in its relationship to their spiritual essence, or soul.

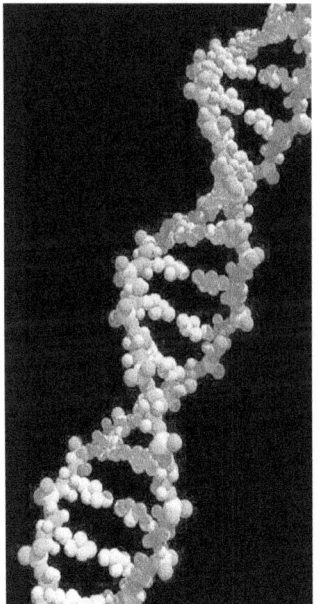

The nervous system routes everything to the brain. It sends signals to the spinal cord, then to the brain. The brain, in turn, communicates that stimuli to your soul and is interpreted as *three-dimensional fractal geometry*. Everything that you taste, touch, see, smell and hear is stored within the brain in this manner. Your DNA has to be dynamic in order to accommodate this feat. Your DNA has three-dimensional fractal geometry encoded *within it*, because it is throughout your entire body.

It also has the master numbers 11, 22, and 33 within it, because your ribs are 11 on each side, your skull has 22 bones in it, and your back has 33 vertebrae.

http://hardtruth.navhost.com/dna.html

The Pineal Gland is the eye of Ra the all Seeing Eye

One explanation for these fractions states that each fraction corresponds to each of the six senses with which we experience our subjective reality. To the usual 5 senses, a sixth sense is added, the sense of kinaesthetic or proprioceptor. **Even if we combine all the experiences of our senses, we cannot comprehend the totality of reality, just as the six fractions added together only total to 63/64, not 1, the symbol of that perfect totality.** The ancient Egyptian belief claims that a being or soul reaches perfect perception of reality only in the next world. This belief is symbolized in the numerical values of the sections of the Eye of Horus."

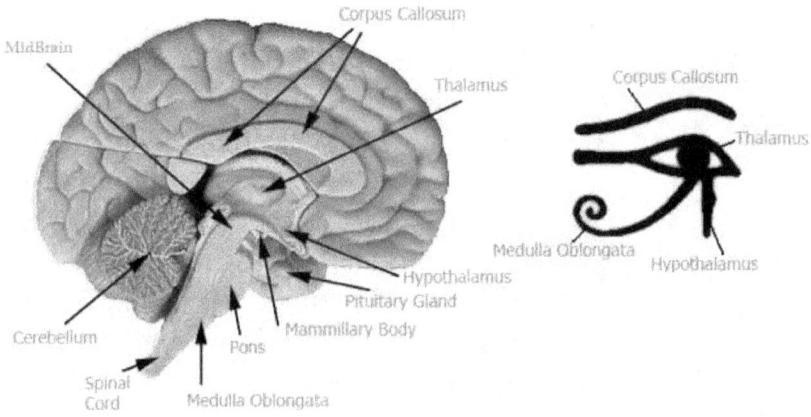

This All-seeing eye was called "**Horus who rules with two eyes.**" His right eye was white and represented the sun; his left eye was black and represented the moon. The Eye of Horus is also known as Wedjat or Oudjat - which means 'sound eye'. It is comprised of the **left lunar eye**, connected with the moon, and the **right solar eye**, the eye of RA, associated with the sun. Here is an ancient equation for the science of eclipses, the wisdom of the Elders, and a call-out for balancing duality -- all wrapped up into one. It is the Holy Grail of Humanity - the Resolution of Duality: Masculine and Feminine, White and Black, Sun and Moon. Same Blood. Dissolved in Love. *Will it ever happen?* That's what we are working on and during this eclipse perhaps more will be revealed to you concerning this primal issue of life on earth (it's a Polarity Planet)

Information was from this link
http://archive.constantcontact.com/fs166/1101472333529/archive/1111662380316.html

Is the Pineal Gland a Crystal Transducer?

The PINEAL GLAND is located within the human brain though its full potential is only beginning to be realized by modern scientists.

The pineal gland was the last endocrine gland to have its function discovered. Its location deep in the brain seemed to indicate its importance. This combination led to its being a "mystery" gland with myth, superstition and even metaphysical theories surrounding its perceived function.

Rene Descartes called the pineal gland the "seat of the soul", believing it is unique in the anatomy of the human brain in being a structure not duplicated on the right and left sides. This observation is not true, however; under a microscope one finds the pineal gland is divided into two fine hemispheres.

The pineal gland is occasionally associated with the sixth chakra (also called Ajna or the third eye chakra in yoga). It is believed by some to be a dormant organ that can be awakened to enable "telepathic" communication.

It is already known to release various chemicals into our body, including a derivative of the feel good chemical serotonin, called melatonin. This hormone affects the modulation of our waking and sleeping patterns, but also affects our sex drive according to the seasons. Though scientists still admit that they still don't yet have a complete picture of the pineal gland's functions.

It is located in the hidden centre of the brain. It is pine cone shaped and no bigger than a raisin. Incredibly, it is actually bioluminescent, so glows within the darkness of the brain as if lit by a tiny light bulb, and has also been found to be sensitive to light. Interestingly enough, the anatomy of the gland actually consists of a Lens, Cornea and Retina Just like our eyeballs. Also, according to scientist Dr Grahame Blackwell, a large number of small crystals have been found in the gland called calcite micro-crystals. They bear a striking resemblance to the calcite crystals in the inner ear that have been shown to exhibit the qualities of an electric field known as piezoelectricity.

If the pineal gland crystals exhibit the same qualities, then this would provide a means whereby an external electromagnetic field might directly influence the brain.

It is already known to release various chemicals into our body, including a derivative of the feel good chemical serotonin, called melatonin. This hormone affects the modulation of our waking and sleeping patterns, but also affects our sex drive according to the seasons.
Though scientists still admit that they still don't yet have a complete picture of the pineal gland's functions. It is located in the hidden centre of the brain. It is pine cone shaped and no bigger than a raisin.

Incredibly, it is actually bioluminescent, so glows within the darkness of the brain as if lit by a tiny light bulb, and has also been found to be sensitive to light. Interestingly enough, the anatomy of the gland actually consists of a lens, corona and retina.

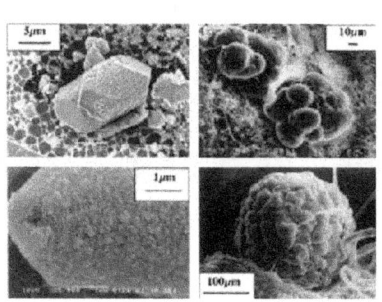 

Left: The Court of the Pine Cone (Vatican City State).
Top Right: The staff of the Pope. **Bottom Right**: The staff of Osiris (Egyptian God)

At close magnification the Calcite Micro-Crystals are visible on the actual gland.

Therefore, it is probably not surprising why over centuries of human history; esoteric groups consider the pineal gland (or All Seeing Eye) to be our built-in wireless transmitter, enabling us to connect to higher frequencies and spiritual worlds.

You can see representations of this pine shaped gland, in the form of a pine cone, across Europe and Egypt.

The Vatican built the court of the pine cone, which is adorned with a large stone pine cone in front of its entrance. It is also found on the staff of the Pope, and the Egyptian god Osiris and in the Sumerian culture too.

This information was from
http://nexusilluminati.blogspot.com.au

The secrets of the carbon atom

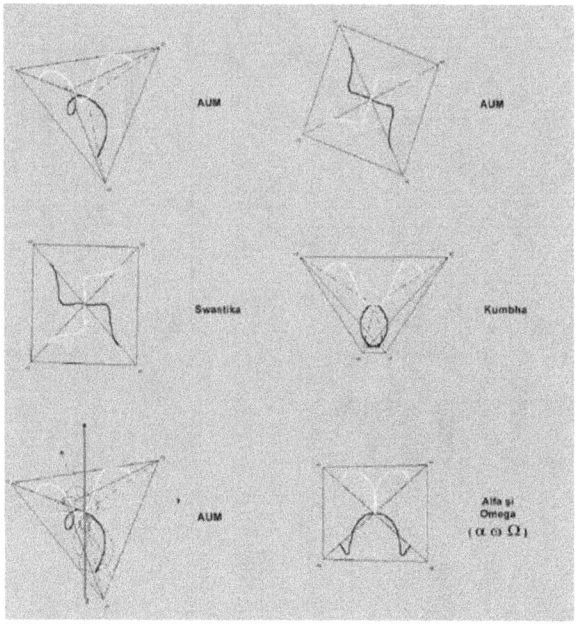

By rotating the Tetrahedron again in relation to the presentation of the Carbon atom you can see the shapes from letters and symbols. You will clearly see the swastika appear and the Alpha and Omega. To see this rotation occur follow the link below.

http://www.themeasuringsystemofthegods.com/Images/Carbon%20atom.gif

Musical proportion of the Fibonacci Sequence

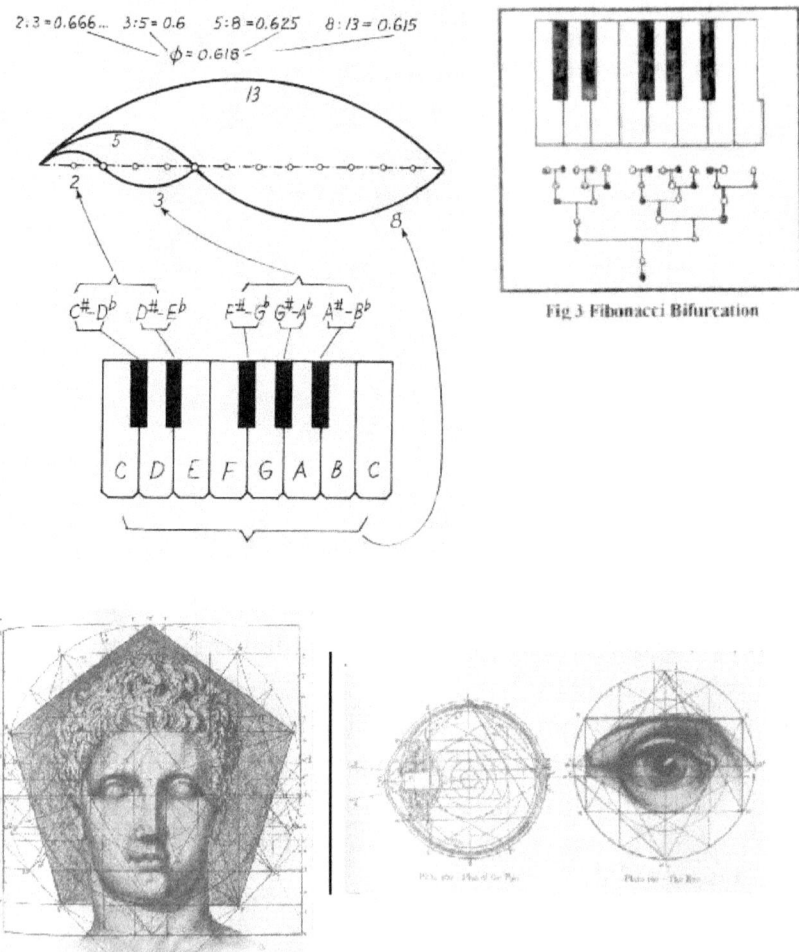

Fig.3 Fibonacci Bifurcation

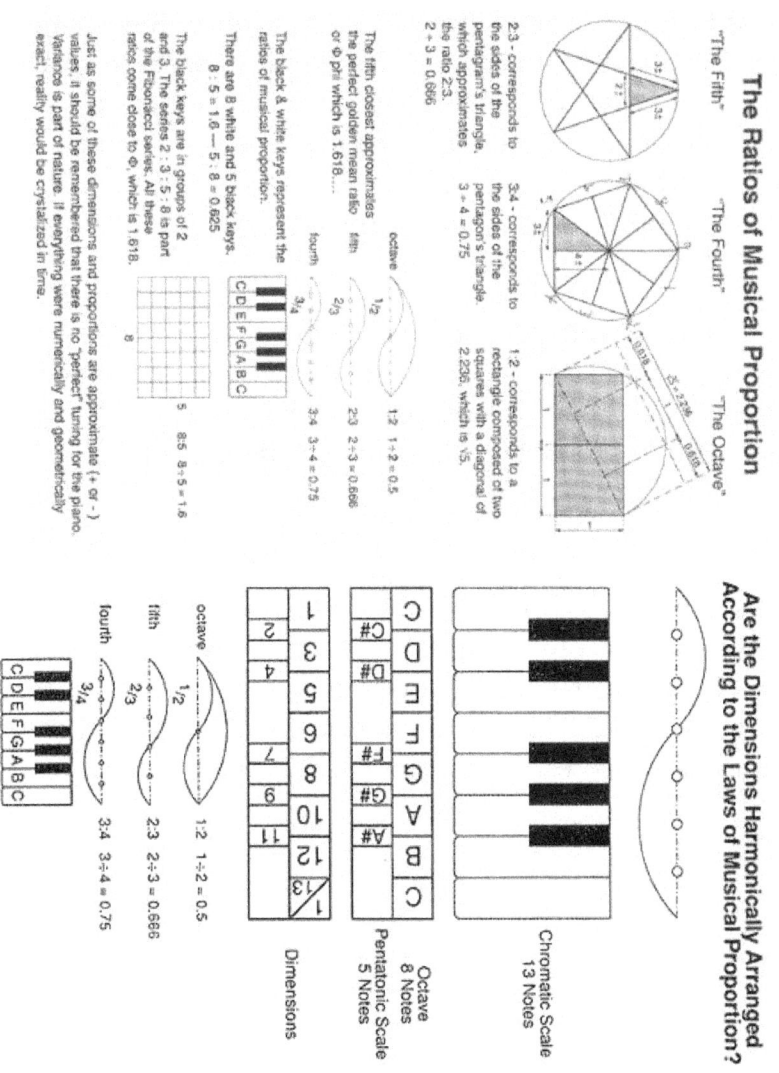

Musical proportions create all the designs in nature and the following images will give you an idea of why people reverse engineer nature. Most of these techniques are not taught in Architecture or engineering and they it should be compulsory for all to know.

These techniques are were used by masters of their craft like Michelangelo and Di Vinci and many others from a bygone era that appreciated perfection, not abstract interpretations of whatever nonsense is going through an "artist" mind.
http://www.miqel.com/jazz_music_heart/vibrational-truth.html

Modern CAD software does not really allow you to create drawings or designs like this and therefore people have moved away from the Compass and Square to an inferior system of design just for the sake of convenience.

The following two books "Harmonic Proportion and Form in Nature, Art and Architecture" By Samuel Colman 1912 are the images above and "The Power of Limits, Proportional Harmonies in Nature, Art and Architecture" Gyorgy Doczi 2005 are the images below. Anyone who is interested in these design principles should purchase a copy.

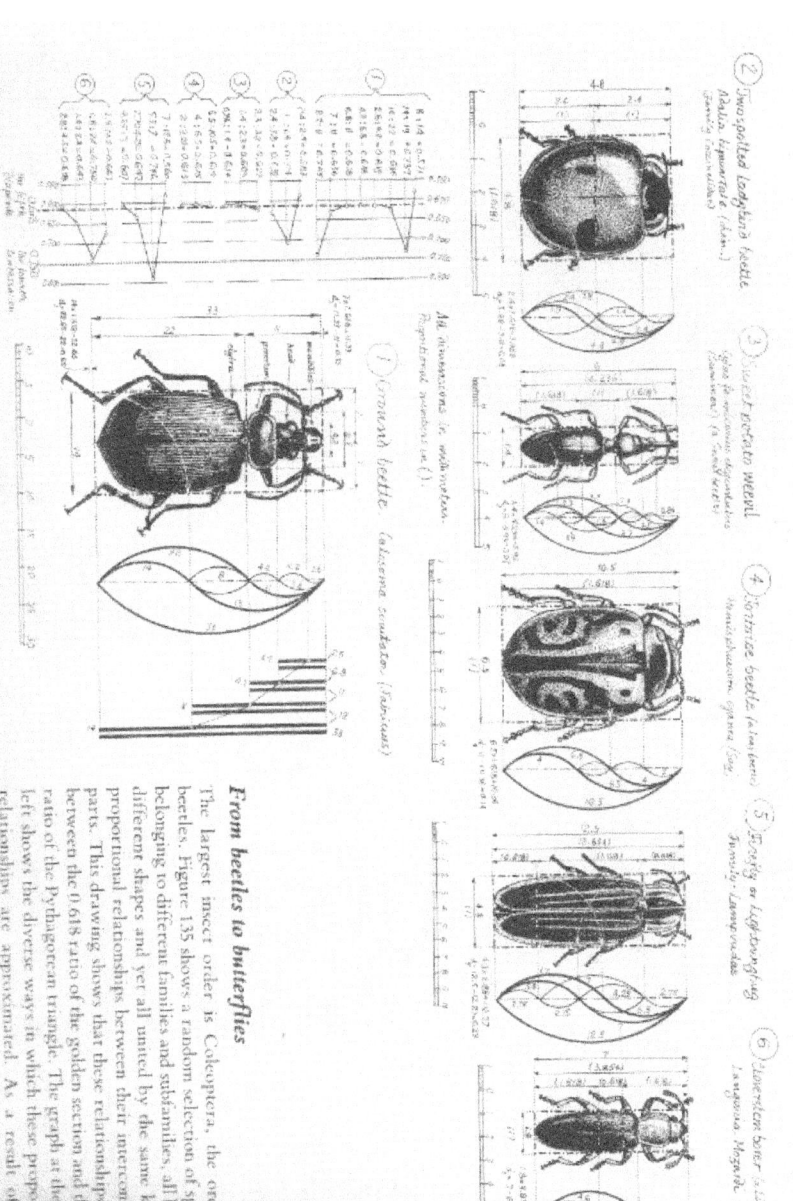

Fig 135. Unity in the diversities of beetle shapes

From beetles to butterflies

The largest insect order is Coleoptera, the order of beetles. Figure 135 shows a random selection of species, belonging to different families and subfamilies, all having different shapes and yet all united by the same kind of proportional relationships between their interconnected parts. This drawing shows that these relationships carry between the 0.618 ratio of the golden section and the 0.5 ratio of the Pythagorean triangle. The graph at the lower left shows the diverse ways in which these proportional relationships are approximated. As a result of these shared relationships, the overall proportions as well as

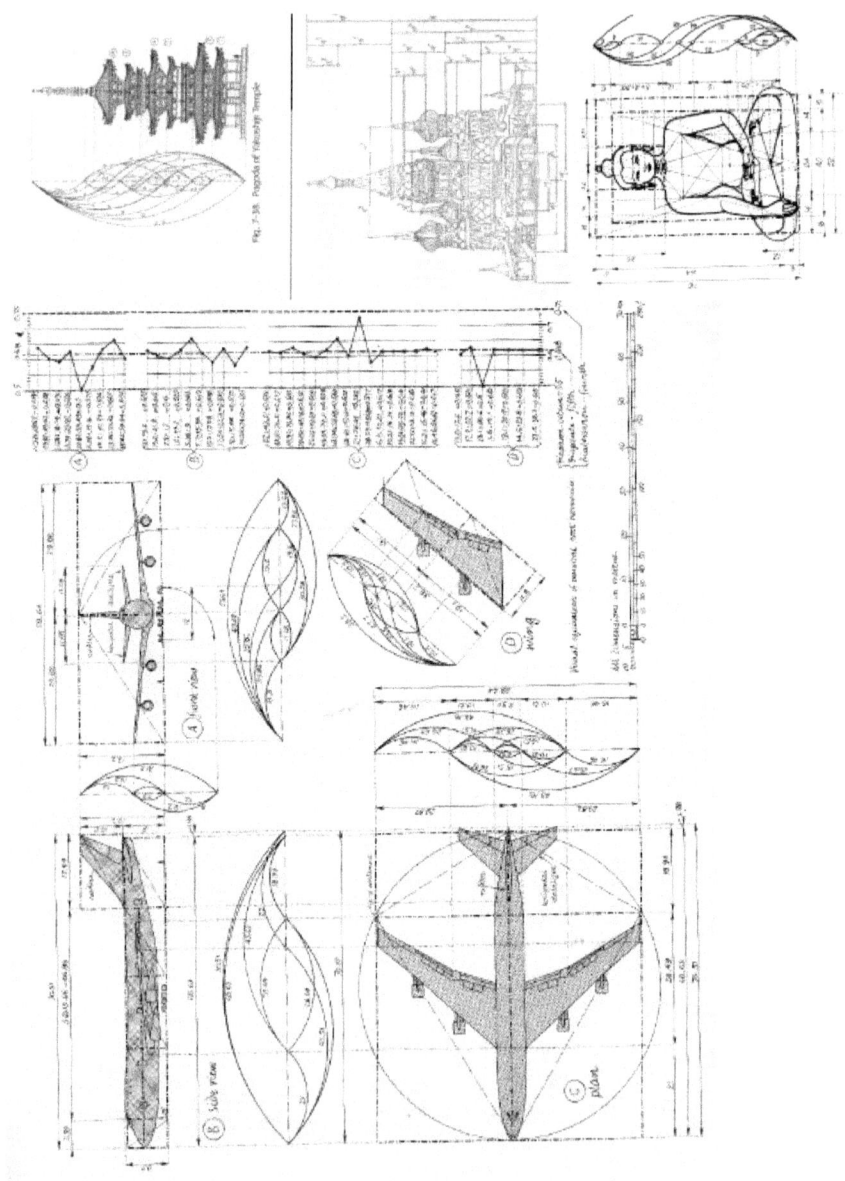

These images are from "Implosion" by Victor Schauberger from Austria

Note the enlarged image of the spiral below and look for 270, 360 degrees, etc. When designing a natural vortex make sure your 270mm and 360 mm coincide with these points to ensure an exact replication of a natural vortex.

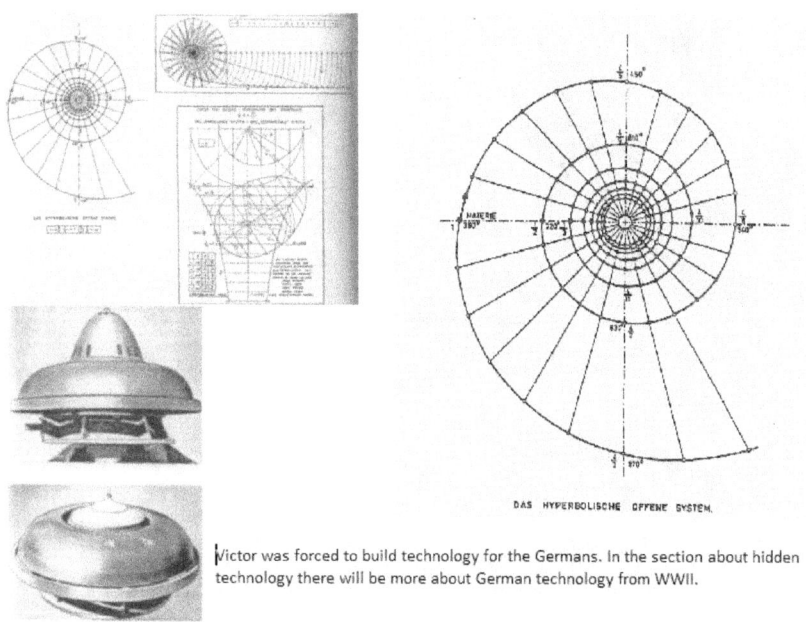

Victor was forced to build technology for the Germans. In the section about hidden technology there will be more about German technology from WWII.

Victor was forced to build technology for the Germans. In the section about hidden technology there will be more about German technology from WWII.

A basic version the golden ratio is used by Advertising Agencies to design company logos and products as it is more appealing to the human eye. Here are two links to show you more examples.

http://www.banskt.com/blog/golden-ratio-in-logo-designs

http://www.phimatrix.com
IRREDUCIBLE COMPLEXITY

Irreducible complexity (IC) shows that certain biological systems are too complex to have evolved from simpler or "less complete" predecessors, through natural selection acting upon a series of advantageous naturally occurring or chance mutations.

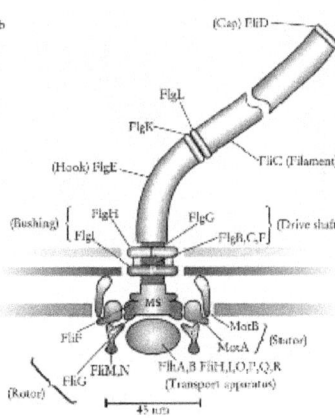

FIGURE 2. BACTERIAL MOTOR AND DRIVE TRAIN. (a) Rotationally averaged reconstruction of electron micrographs of purified hook-basal bodies. The rings seen in the image and labeled in the schematic diagram (b) are the L ring, P ring, MS ring, and C ring. (Digital print courtesy of David DeRosier, Brandeis University.)

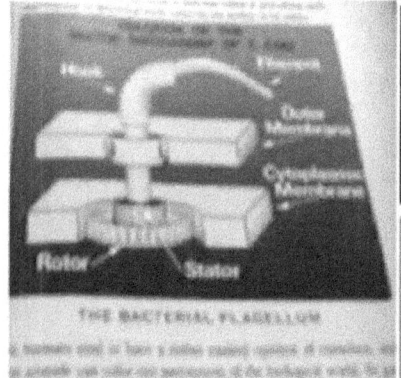
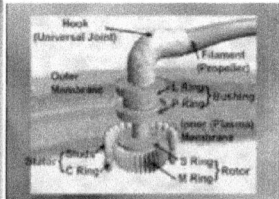

- "What is 'design'? Design is simply **the *purposeful* arrangement of parts**."

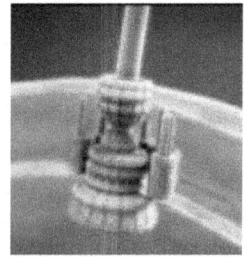

Biochemistry professor Michael Behe, the originator of the term *irreducible complexity*, and Author of "Darwin's Black Box" defines an irreducibly complex system as one "composed of several well-matched, interacting parts that contribute to the basic function, wherein the removal of any one of the parts causes the system to effectively cease functioning"

DeRosier, D. J. (1998). The turn of the screw: the bacterial flagellar motor. *Cell* **93**, 17-20.

- "More so than other motors, the flagellum resembles a machine **designed by a human**."

We're told that life evolved from the very simple to the highly complex, which would make sense if it were true.

The truth, however, is that even the simplest of life forms are highly sophisticated pieces of biological machinery.

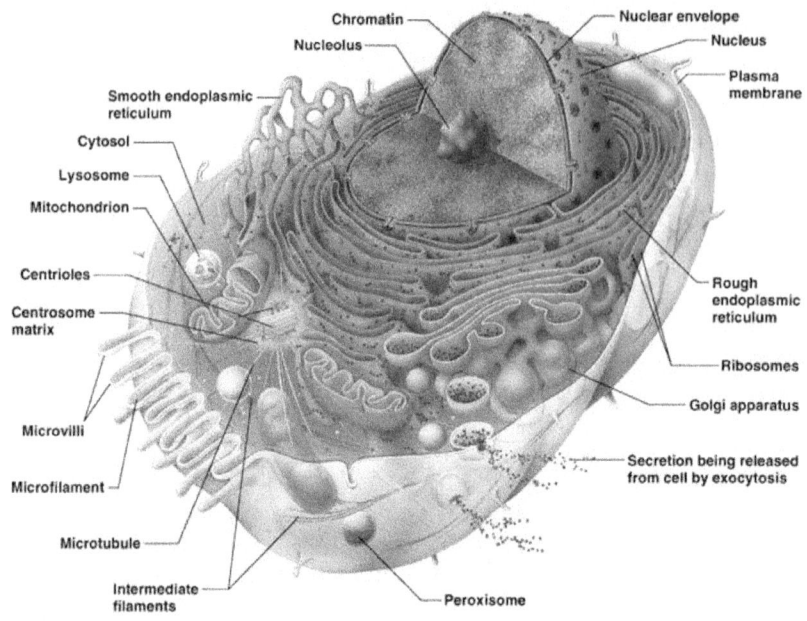

Every living cell contains DNA, the program which describes every aspect of the organism's physical design and life functions, including:

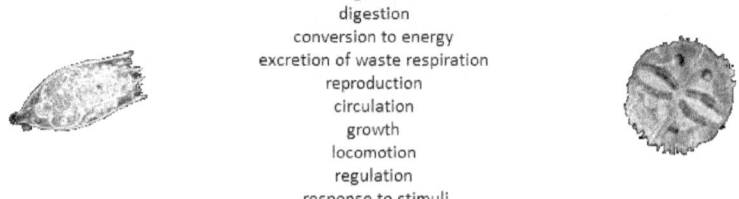

ingestion
digestion
conversion to energy
excretion of waste respiration
reproduction
circulation
growth
locomotion
regulation
response to stimuli

A cell would not really be "alive" or survive unless all life functions became operable at once, so how did even a "simple" single-celled organism "evolve"? We're told by scientists and academicians that finding or creating a simple molecule of amino acid is evidence of "life," yet DNA of even a simple E coli bacteria cell contains 4.6 million base pairs, or 9.2 million bits of information.

Does evolution theory offer any evidence or even a truly plausible explanation as to how even a single bacteria cell came into being?
http://evolutionoftruth.com/evo/evogene.htm

http://evolutionoftruth.com/evo/evocmplx.htm
Creation over Evolution

The Vedas and the worship of Krishna and the Supreme is the world's oldest religion consisting of a real science used in secret today.

The Vedic Metric and Vedic Physics which includes all the principles of Harmonic Proportion and Form, Sympathetic Vibration, magic squares and cubes for different time's places and dimensions all prove Intelligent Design.

Everything you see in natures was created simultaneously, purposely designed to perform a function. With all this information in front of you about Euclidian, Fractal geometry and the mathematical musical formation of everything you see within your mind, to say that evolution is a plausible explanation is absolutely insanity.

I have showed you that physicists have reverse engineered their metrics from the ancient RG Veda, which are more than 5,000 years old. How old is the Bhagavad-gita?
Now we have just passed through five thousand years of the Kali-yuga, which lasts 432,000 years. Before this, there was the Dvapara-yuga (800,000 years), and before that, there was Treta-yuga (1,200,000 years). Thus, some 2,005,000 years ago, Manu spoke the Bhagavad-gita to his disciple and son Maharaja Iksvaku, the king of this planet earth.

The age of the current Manu is calculated to last some 305,300,000 years, of which 120,400,000 have passed. Accepting that before the birth of Manu, the Gita was spoken by the Lord to His disciple, the sun-god Vivasvan, a rough estimate is that the Gita was spoken at least 120,400,000 years ago; and in human society it has been extant for two million years.

Ancient Astronauts & Suppressed Information

This chapter as completed using a common sense approach to what many intelligent people believe to be the true history of the world. I do not hold any faith in the ability of many of our educational institutes to provide a valid explanation of the world around us. As you would have already read in this book there is more going on than you are being told. History is written by the victors. Most people can't imagine that a spaceship could be the size of a planet, the design and construction of this would be inconceivable. Well it is possible using the science of creation and it has happened. We are not different races of humans but different species of humanoid aliens on one world.

Iapetus is not origin of man but part of the history of the human race.

http://www.nasa.gov/mission_pages/cassini/multimedia/Cassini_Multimedia_Collection%28Search_Agent%29_archive_2.html

03.29.10
Flying by the "Death Star" Moon

NASA calls this the "death star" moon.

http://www.nasa.gov/mission_pages/cassini/multimedia/pia12570.html

The information on this page is from this website but it does not exist anymore.

http://www.conspiracynewsnet.com/shadow.html

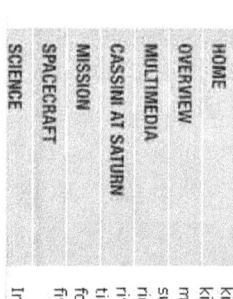

Cassini-Huygens
MISSION TO SATURN & TITAN

For News Media
- Introduction
- News Releases
- Significant Events
- Features
- Mailing List Sign-up
- Fact Sheets
- Press Kits
- Contact Us

For Educators
For Planetariums & More
For Kids
HOME
OVERVIEW
MULTIMEDIA
CASSINI AT SATURN
MISSION
SPACECRAFT
SCIENCE

NEWS - News Releases - 2005

Saturn's Moon Iapetus Shows a Bulging Waistline

January 7, 2005
(Source: Jet Propulsion Laboratory)

Images returned by NASA's Cassini spacecraft cameras during a New Year's Eve flyby of Saturn's moon Iapetus (eye-APP-eh-tuss) show startling surface features that are fueling heated scientific discussions about their origin.

One of these features is a long narrow ridge that lies almost exactly on the equator of Iapetus, bisects its entire dark hemisphere and reaches 20 kilometers (12 miles) high. It extends over 1,300 kilometers (808 miles) from side to side, along its midsection. No other moon in the solar system has such a striking geological feature. In places, the ridge is comprised of mountains. In height, they rival Olympus Mons on Mars, approximately three times the height of Mt. Everest, which is surprising for such a small body as Iapetus. Mars is nearly five times the size of Iapetus.

Images from the flyby are available at http://saturn.jpl.nasa.gov.

Iapetus in 3D
More Iapetus Images

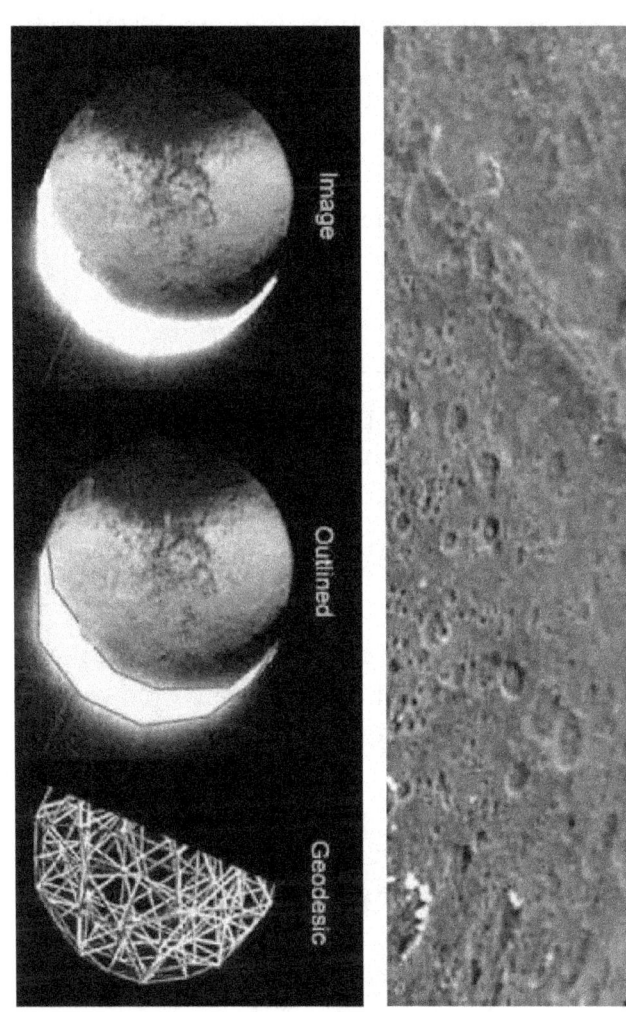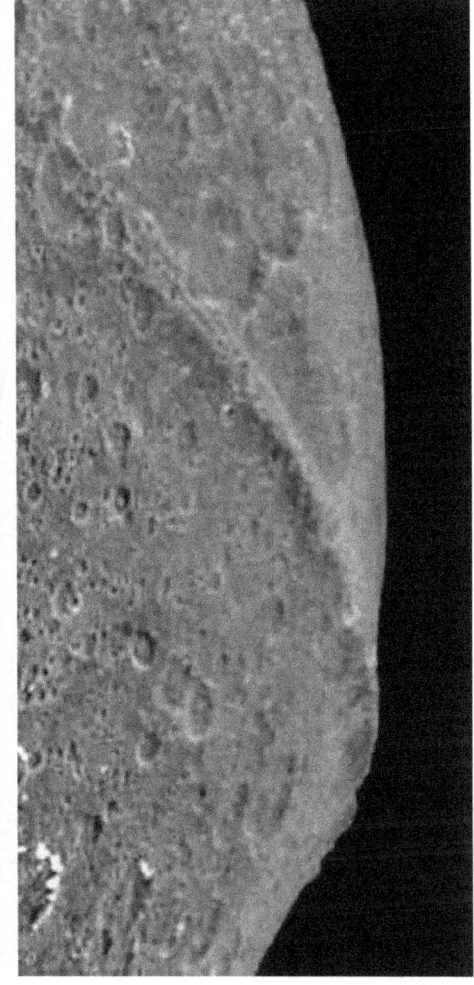

The ancient civilizations that once occupied western Iran to the eastern Mediterranean coast, as well as Egypt and the Saudi peninsula left astonishing depictions and an extensive written archaeological record of their visitation by 'sky gods,' or the ANNUNAKI (those who from heaven to earth came).

Many books and articles have been written on this subject, and I will only attempt to give an overview of the general information about them here, as it may relate to this article. According to ancient records and artifacts, beings visited Earth from a planet called NIBIRU. Their world, the visitors claimed, is dragged along with a smaller brown dwarf star that orbits our sun- an enormous orbit that takes almost 4,000 years to complete.

These ancient records describe this dwarf star, dragging along its planets, as periodically passing near enough to our solar system for them to send ships to earth, where they would engage in accumulation of gold and other metals through mining operations using bituminous (oil/petroleum based) products as fuel to refine ores. This would occur over a period of several hundred years. The visitors would send these materials into low earth orbit, eventually returning with them to their ship, which would remain in orbit between Mars and Jupiter.

These visitors claimed they had been coming to Earth for the last 450,000 years, and that in the distant past they had developed a 'worker class' through genetic engineering involving themselves and early hominids that were in fact the first humans. Over time, these beings claimed, their periodic visitations (every 3600-3800 years) led to the formation of human high civilization, including the development of government, agriculture, written language, and various sciences. Concepts relating to religions and deities, along with depictions of winged objects in the ancient sky that is found in the region's archaeological record, are apparently attempts to emulate or record and illustrate these visitations.

Many of these visitors remained behind after their companions left, and were credited by these cultures with teaching men to quarry and build with massive blocks and segments of stone as well as helping humans to design and construct the ancient cities, temples, and agricultural systems.

According to Sumerian records, the Annunaki claimed that as their planet approached our system they would arrive from their world on ships, each containing fifty occupants, materials and supplies.

These ships would then enter our solar system, assuming an orbit passing between Jupiter and Mars. They would remain there for several hundred years, as teams sent to Earth gathered gold and other metals from southern Africa and the region between the Tigris/Euphrates Rivers, sending them back to the orbiting ship.

Others gathered minerals and ores from the asteroid belt, which they called RAKKIS (the hammered bracelet). As they completed operations, their ship would leave its orbit and approach Saturn. Using Saturn's enormous gravitational pull to assist their acceleration they would then 'slingshot', returning to their own system which would be completing the nearest part of its 4,000 year orbit and heading back out into deep space.

On January 9, 1880, a report to British Royal Astronomical Society by R.H.M. Bosanquet and A.H. Sayce described a clay tablet circa 3300 BC, discovered in the ruins of the Royal Library in Nineveh.

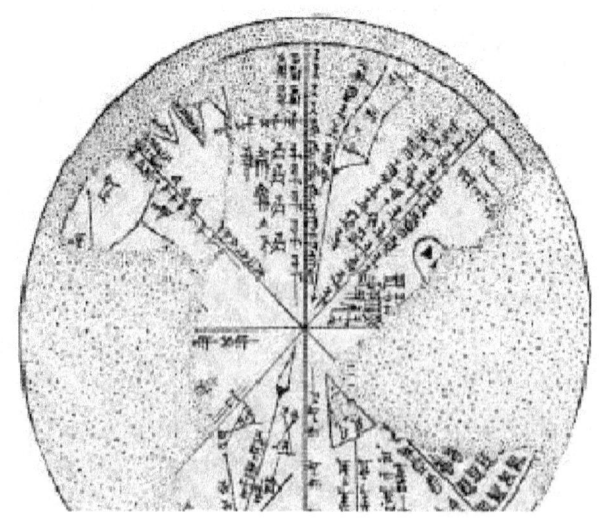

This circular tablet, divided evenly into eight equal segments, was described as a 'planisphere' or map of the heavens. Several segments shapes are associated with higher mathematics, and one section has a clear diagram of the constellation of Orion. In a very striking segment that is almost completely intact, there are two triangular shapes, with a row of dark circular shapes stretching between them. Under this row, written in cuneiform is the phrase "the deity EN.LIL goes by these planets". In this same segment, a large triangular shape with parallel lines is visible. This was a commonly used pictograph in the Near Eastern region, used to indicate a ruler's domain, or point of origin when a ruler undertook a journey.

This is connected with an angular line to another smaller triangle, with the names of two celestial bodies written above it. These cuneiform terms translate as DIL.GAN (first marker) and APIN (where the course is set). According to the scholar Strassmeier, Apin represents Mars, so DIL.GAN (first marker) would be massive, easily seen Jupiter. These seem to indicate an orbital position of their ship in our solar system between Jupiter and Mars.

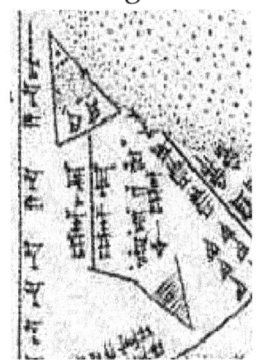

Stretching between these two triangles there is a row of dots, indicating planets. The Annunaki considered the earth's moon to also be a planet in its own right, so a count of the planets outwards from the sun would put their line of travel as passing Saturn on their way into or out of our solar system.

In the lower triangle in this segment there is also written in cuneiform the term SHUUT.IL.EN.LIL, or 'the way of Enlil'. This was another very common term used in the region, this one indicating the sky or heavens. The sky. In the lower part of this section, we also see a group of four dark spots, carefully arranged in a pattern with one leading centred spot. The positions of these aligned dark spots directly correspond with the locations of the ruins of the ancient cities of Larsa, Bad-Tibira, and Lagash in southern Iraq. Nippur, or their landing site on Earth, is indicated by the dark spot in the lower corner of the triangle, and is also located in this exact spot in relation to the other three ancient cities in the Tigris/Euphrates river valley.

This section of the planisphere describes the navigational route of these beings through our solar system past indicated planets to an orbit from which smaller craft (SHEM) were flown to a landing site on Earth. It is very similar to the imagery engraved on a plaque attached to the NASA Pioneer 10 probe that left earth for deep space in February 1971 contains a planet map, with the path of the departing Pioneer 10 probe as it heads out of our system.

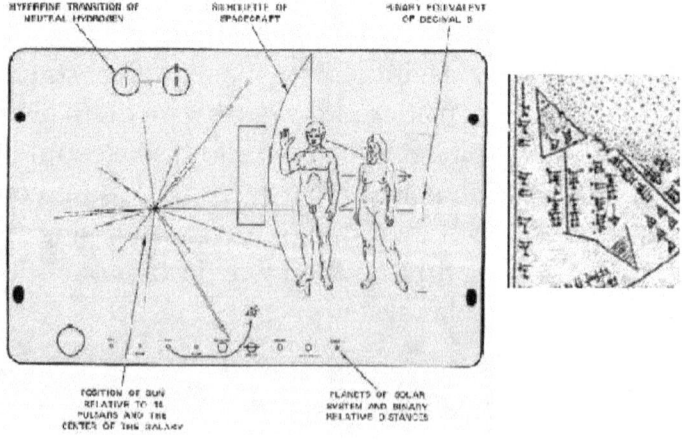

Saturn's enormous gravitational and electromagnetic fields were regarded with great fear and awe by the Annunaki - they called it TAR.GALLU, or 'the great destroyer'. Apparently this was because of an incident in their distant past when a ship with fifty occupants was lost on a mission to the inner solar system. Possibly through trajectory or flight path miscalculations, the story describes a craft becoming trapped in a large orbit around Saturn, hopelessly dooming its occupants from any timely rescue. A Sumerian cuneiform text, published in 1912, describes this incredible disaster as an Annunaki named MARDUK rushes to his father, EA, with terrible news.

"It has created itself like a weapon... It (the ship) has charged forward into oblivion! The Annunaki who number fifty it has seized! It has seized SHU.SAR(planet chaser) in its grip of death!"

The 2004/2005 Cassini/Huygens probe's flyby of Saturn revealed unprecedented photos of many bodies in the region, including close-ups of Saturn's moon Titan and its gaseous atmosphere. Only briefly mentioned in the media and in the official NASA releases, however, was the fact that a correction in the flight path of the probe had sent it within photographic range of Iapetus, a body that orbits Saturn in a very large and unusually inclined orbit- far wider than any other object circling Saturn. Scientists believe this extremely remote orbiting object was most likely 'trapped' or pulled into orbit instead of originating as part of Saturn's moon and ring system.

NASA was baffled as readings from sophisticated equipment aboard the Cassini/Huygens probe indicated that the object was very likely hollow. As the first photos of Iapetus began to appear, NASA and the scientific world were shocked to see that this was clearly an artificial object, manufactured and with clear indications of intelligent design.

The object showed advanced yet very ancient construction and technology, on a massive scale. It became clear that this object was an enormous, very ancient abandoned space craft, very similar in design to the spacecraft Death Star from the George Lucas science fiction film Star Wars.

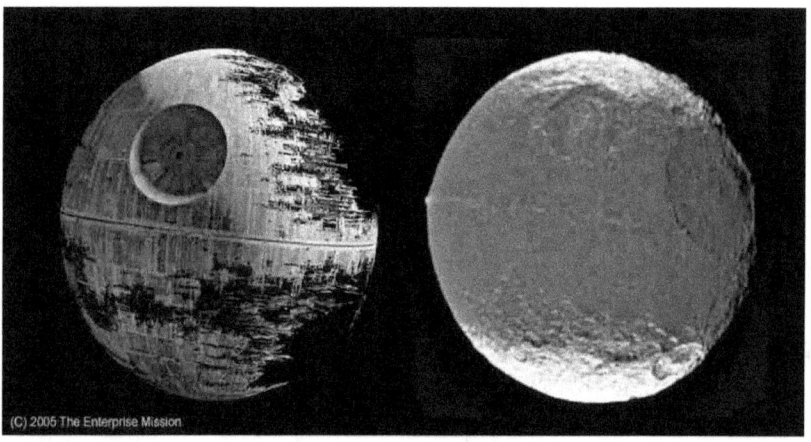

This revelation was quickly suppressed, and official NASA releases only indicated that there were strange features visible on the object's surface.

Iapetus, shaped like a peanut, has a perfectly straight 60,000 foot high bank of surface material that bisects its centre, running the entire visible length of the object. Obviously an artificial construct, this is apparently to protect a seam along this part of the ship. The ridge also has an internal structure of three beams or tubes that run parallel and are inside the embankment.

This object also has what appear to be two equally sized and spaced exhaust ports on one extremity. These are surrounded by darkened areas, possibly from heat or exhaust residue.

In this photo from the probe, the two exhaust ports, equal in size and shape, are seen on the rear of the craft. Note the darkened regions on the surface, which emanate from these ports. These areas are consistent with exhaust burns and residue from a propulsion system.

Iapetus also has a very large number of craters that appear to be symmetrical, mostly hexagons. These seem to show an underlying tetrahedral framework, with ports at various places, with overlying layers of rocky strata. The largest and most prevalent surface feature, other than the ridge straddling its centre, is a large hexagonal portal or bay, with clearly angular and symmetrical sides.

Along the edges of this same area, edge collapse and degradation have exposed massive structures and engineered seams and angled remnants, protruding from eroded and exposed edges.
From ancient mines in southern Africa, villagers have found small finely manufactured spheroids of unknown origin and made from milled metals hundreds of thousands of years ago. This is the same area where the Sumerian accounts claim the Annunaki were engaged in mining operations for metals.

These metallic objects are precise models of the ship that is now trapped around Saturn. There have been two distinct types of this object discovered, one of which models the three covered ridges found straddling the central region of Iapetus.

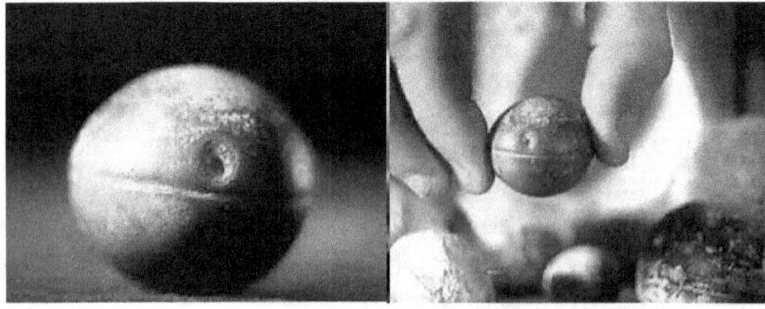

Unfortunately there has been little mention of this unprecedented discovery in the mainstream or alternative media, and the scientific community is strangely silent on the subject. Clear evidence of artificial design and manufacture of an extra-terrestrial object in our solar system, along with correlating physical and archaeological evidence of its existence should be viewed as one of the most significant scientific finds in the history of mankind. This discovery, however, has been largely ignored.
In 1959, NASA issued a document called the Brookings Report, which warned the US Government that any scientific evidence of extra-terrestrial intelligence "could be destabilizing to terrestrial governmental institutions...if not the future of civilization." Such a policy may possibly explain the deafening silence from NASA, the world scientific authorities, and governments regarding this monumental scientific discovery.

We have been invaded but this invasion was not with guns and bombs but with advanced technologies being used over a long period of time. Advanced civilizations invading our world would not require simple weapons but only to use the control of our media, banking and political systems against us. The most successful military campaign is one in which apparently no shots are fired and the conquered populace gladly welcomes their enslavers. We have been mentally and socially programmed and conditioned to accept our invaders as saviors, not a conquering force.

Richard Dolan, May 3, 2013, part of his final statement to the Panel for the Citizen Hearing on Disclosure, which ran from April 29 to May 3, 2013. The panel members were all former members of the U.S. House of Representatives and Senate.

I've often felt that disclosure on the matter of UFOs and possible ETs is a paradox. It is impossible, but it is inevitable. Impossible because there is no political motivation for it. Period.
Inevitable, however, because our leaders are not the only factor in the equation. There are the other beings, after all. But mainly, there's us. The People. Who are going through the greatest social, cultural, and especially technological transformation in the history of humanity.
 In fact, we are the game changers. Someday, and it won't be too long in the future, something is going to force someone's hand. It could be a major sighting, a major leak, something. Something that can no longer be denied. After all, we are fast approaching what experts in artificial intelligence call the Singularity, when computing intelligence exceeds our own. In such a future, can we really think we will still be stuck in neutral on this issue?

Mayan King Pakal and His 'Spacecraft'

These civilizations were extremely advanced.

Do some research into "black Knight" A 10,000 year old man made satellite orbiting around the solar regions of our planet. **The three images to the right are examples of Ancient circuit board technology for computer systems.**

http://onedotatatime.blogspot.com.au/2009/06/ancient-technology.html

Ancient lighting system recreated.

Below the electron tube was designed on the same principles as the ancient light

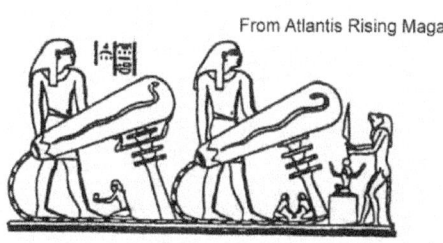 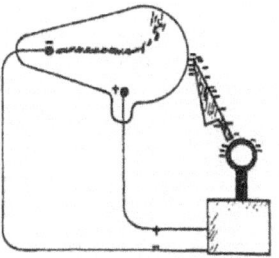

From an engraving on the wall at the temple of Hathor in Dendera, Egypt. The accompanying hieroglyphs describe the serpents as 'seref', or 'to glow'.

Schematic diagram of a Crooke's electron tube, the forerunner of the modern TV tube.

Between the avenue of Sphinxes stood two generators, A and B. A is sending and B is receiving. Between these two antennas a solitron or vortex field is generated. It is on the field that the heavy building blocks were carried.

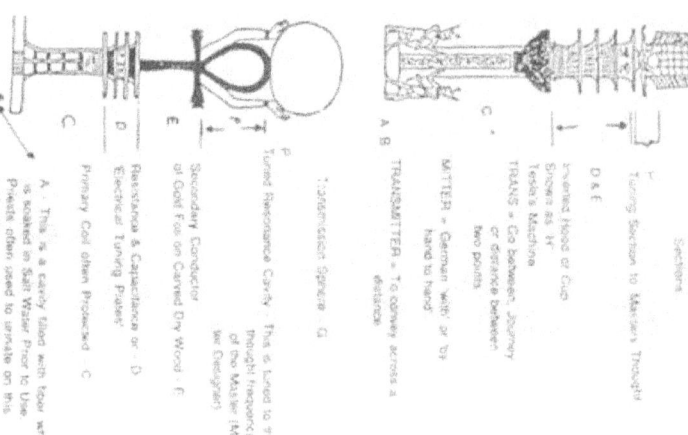

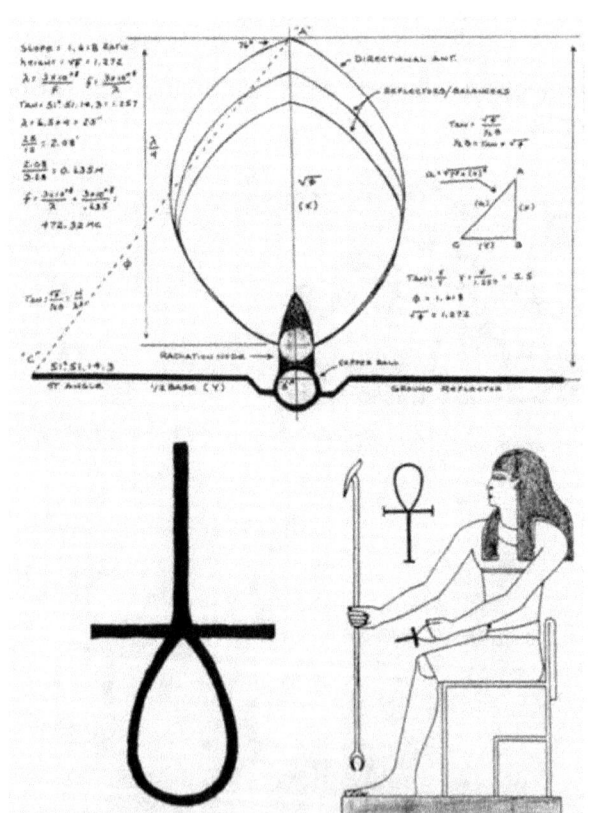

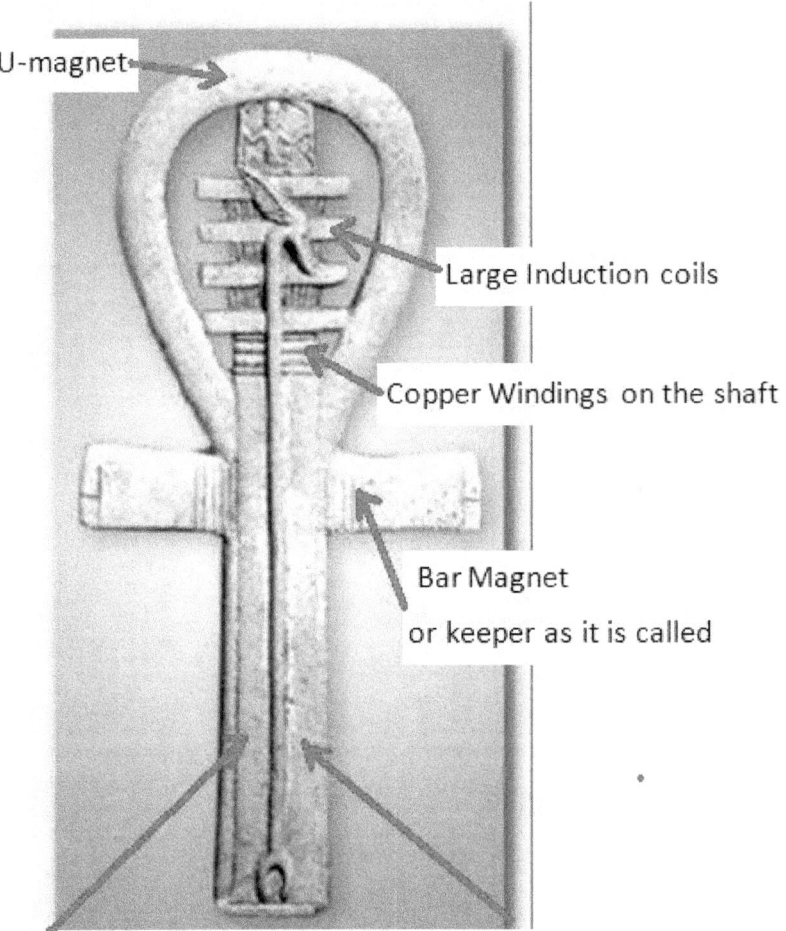

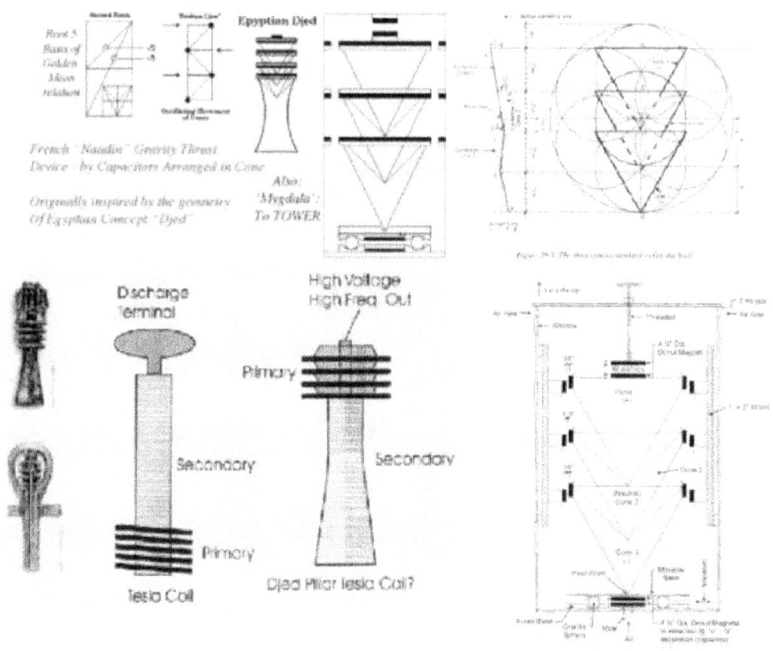

Elongated Skulls of King Tut and Family show they have been here for thousands of years.

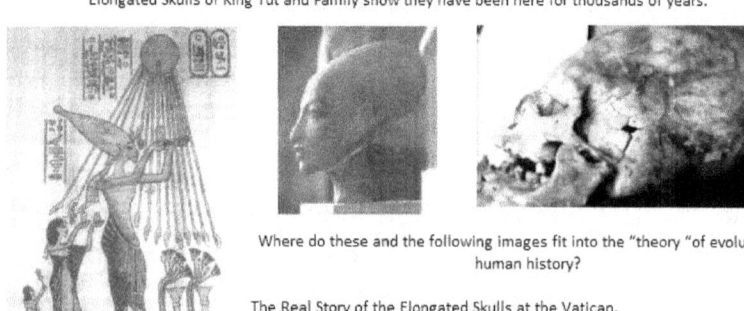

Where do these and the following images fit into the "theory" of evolution or human history?

The Real Story of the Elongated Skulls at the Vatican.

The report of these elongated skulls is years old at this stage but it does make you wonder what the Vatican knows and when did they know it.

Where do these and the following images fit into the "theory" of evolution or human history?

The Real Story of the Elongated Skulls at the Vatican.

The report of these elongated skulls is years old at this stage but it does make you wonder what the Vatican knows and when did they know it.

"Whatever those remains represent, there's a reason why the good fathers buried them there to be forgotten." Alien Remains Found at The Vatican. Pope Told To Remain Silent. Many alien skulls were found during a restoration project at the Vatican Library. Rendition of the appearance of the excavated skull when alive.

THE VATICAN. SPECIAL REPORT. Skeletal remains resembling 'space aliens' have been excavated from the basement floor of a centuries old vault under the Vatican Library.

The Library had been undergoing a major restoration to its' underground vaults, of which many, CN has learned, still contained dirt floors that have not felt a human foot in over 500 years. The above picture of a restoration worker holding an alien skull was obtained by CN through an anonymous carrier. The Vatican Military has closed all entry to the Library. Although a Vatican spokesperson was quoted expressing his awe and excitement concerning the find; he later disavowed any knowledge of the comment. Press sources have confirmed that an unprecedented command of silence has been enacted by Pope John Paul upon the request of several world security organizations.

A world security agency meets with the Pope.

UFO organizations are already calling this "the great world government cover-up." Mary Peterson, chairperson for SART (Space Aliens are Real and a Threat), states her opinion "what lies beneath the Vatican floors belongs in our Universities to be studied. This only proves the theory that world government has known of aliens all along. And the Vatican is essentially its own country with its' own government and military." The 'fear of God' has already been put into the hearts of many of the faithful. Sister Judy Mebosa, chef at St.

John's Cathedral in London, stated "it's the mark that the end is near. Whatever those remains represent, there's a reason why the good fathers buried them there to be forgotten. And there's a reason why God has allowed them to be found."

Michael Maregski, Lutheran pastor of Our Holy Father in Rome, is quoted as saying "the ufologists have been speculating for many years that aliens may have been portrayed as angels to our ancestors in antiquity. If this speculation turns out to be proven true; why would angels be buried at the Vatican only to be forgotten?" Father Edward Muldoon, of St. Jude's in Rome, answered Pastor Michael Maregski's question with "maybe they are angels, but remember, there were two groups of angels."
See more at:

http://www.educatinghumanity.com/2013/03/elongated-alien-skulls-at-vatican-the-real-story.html#sthash.D5wu421f.dpuf

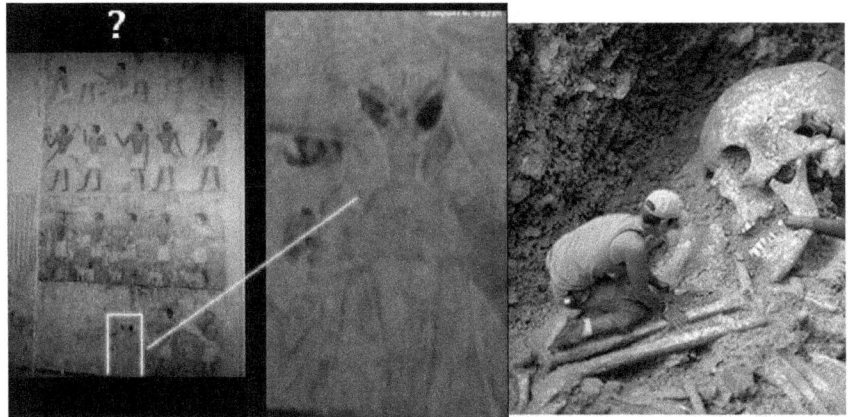

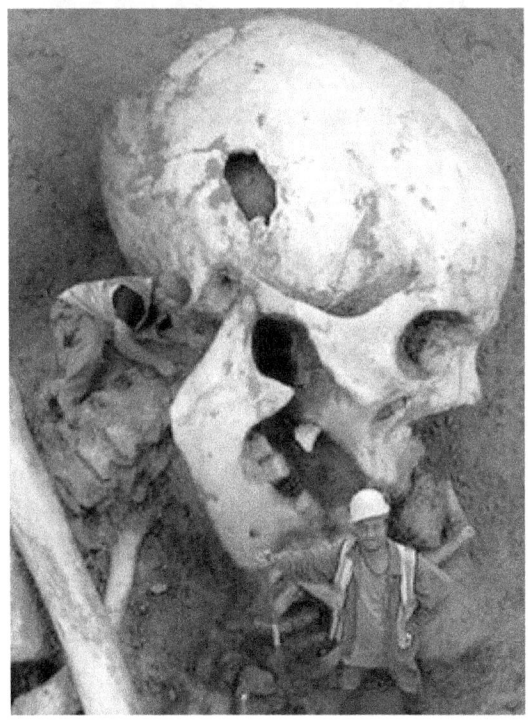

You can see the grants mentioned in the bible, The Nephilim are part of human history and have been mentioned in many cultures and found all around the world. Some have been found in archaeological digs and others found in burial mounds.

THESE SKELETAL FIGURES REPRESENT "JUST A FEW" GIANT HUMAN REMAINS, UNEARTHED AND DOCUMENTED IN HISTORICAL RECORDS, ALONG WITH THE HISTORICAL ACOUNTS OF GOLIATH (who had 3 brothers as big as he), OG King of Bashan, whos bed was 13.5' long and Maximinus Thrax, a Caeser of Rome.

6'
Present day Man

15'
S/E Turkey late 1950s

8'6"
Maximinus Thrax CAESER OF ROME 235-238 AD

10'6"
GOLIATH 1 SAM 17:4 1010 BC

12'
OG King of Bashan Deut 3:11 1400 BC

19'6"
1577 AD Under an Oak tree in the canton of Lucerne

23'
1456 AD France beside a river in Valence

25'6"
1613 AD France, near the Castle of Chaumont. Nearly a complete Skeleton

36'
650 BC - 640 AD Carthaginians uncovered two this size. An earthquake in Cimmorian Bosphorus uncovered one more.

Image from http://extremelifechanger.com

We know they existed, we know where they came from so why are we not taught about them. We know ancient cultures were highly advanced even more so than we are today so why hide it. Surely intelligent people can see we have been deceived. There are web sites that expose artifacts that do not fit into the current version of history taught in universities.

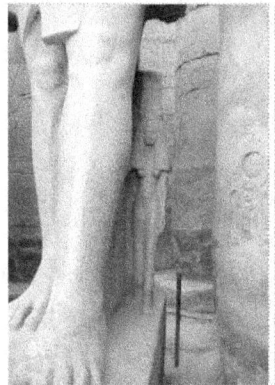 

The ticky tacky world of academia have created different theories and modified history to tell a story that if the truth were known is an act of fraud.

If you charge a fee to a student and the information in that subject was supposed to be gathered through irrefutable peer group analysis and they are aware that the information not factual, that is fraud.

University lecturers are still teaching the same old crusty outdated information they have taught for years because if the content is changed the previous graduates' degrees would be no longer valid. So I will show to you how a degree in history, physics, chemistry are useless because of the information Business degrees are about the only subject a university should be teaching but if you look at the state of the world's economy that is questionable too. I will show you the fraud behind the banking system in the next chapter.

CAVE FOUND IN INDIANA HINTS AGE-OLD RACE

Giant Skeletons and Metals Strange to America Seen In Ancient Sepulchre

I believe the book on the left contains more than enough evidence to prove they existed.

HUGE SKELETON UNEARTHED

Indiana Produces Bones of Man Believed to Have Been Mound Builder.

Indianapolis, Ind.—The complete skeleton of one of Indiana's oldest inhabitants, said by Dr W. N. Logan, state

This hieroglyph would be one of the most fascinating there is to see involving ancient transportation. The interpretation of what they could be seems more than plausible given the advancement of their technology.

Now I am Become Death, the Destroyer of Worlds

Bhagavad-gita (As It IS). These are the words Julius Robert Oppenheimer the American theoretical physicist and professor of physics at the University of California quoted when the first Atomic Bomb was detonated.

"We knew the world would not be the same. A few people laughed, a few people cried. Most people were silent. I remembered the line from the Hindu scripture, the Bhagavad-Gita; Vishnu is trying to persuade the Prince that he should do his duty, and to impress him, takes on his multi-armed form and says, 'Now I am become Death, the destroyer of worlds.' I suppose we all thought that, one way or another."

- J. Robert Oppenheimer (http://www.goodreads.com/author/show/308544.J_Robert_Oppenheimer)

https://www.youtube.com/watch?v=26YLehuMydo
Oppenheimer also claimed that, **"access to the Vedas is the greatest privilege this century may claim over all previous centuries**

While he was giving a lecture at Rochester University, during the question and answer period a student asked a question to which Oppenheimer gave a strangely qualified answer:

Student: *"Was the bomb exploded at Alamogordo during the Manhattan Project the first one to be detonated?*

Dr. Oppenheimer: *"Well — yes. In modern times, of course.*

Many suggest that Oppenheimer was referring to the Brahmāstra weapon mentioned in the Mahabharata. His appreciation of the Hindu culture didn't stop there he always gave the book (Bhagavad Gita) as a present to his friends and kept a copy on the shelf next to his desk. There are many examples of ancient atomic warfare if people wish to investigate this further but I have seen enough of man's inhumanity to mankind to be thoughtful enough not to show you more examples of it.

Battle of hero Adwattan -Mahabharata
The Agneya weapon, a "blazing missile of smokeless fire" is unleashed; dense arrows of flame, like a great shower, issued forth upon creation, encompassing the enemy... A thick gloom swiftly settled upon the Pandava hosts. All points of the compass were lost in darkness. Fierce winds began to blow.

Clouds roared upward, showering dust and gravel... the very elements seemed disturbed. The sun seemed to waver in the heavens. The earth shook, scorched by the terrible violent heat of this weapon. Elephants burst into flame and others ran back and forth in frenzy... over a vast area, other animals crumpled to the ground and died. From all points of the compass the arrows of flame rained continuously and fiercely.

Gurkha, flying in his swift and powerful Vimana, hurled against the three cities of the Vrishnis and Andhakas a single projectile charged with all the power of the universe. An incandescent column of smoke and fire, as brilliant as ten thousand suns, rose in its entire splendour. It was the unknown weapon, the iron thunderbolt, a gigantic messenger of death which reduced to ashes the entire race of Vrishnis and Andhakas.

It was as if the elements had been unleashed. The sun spun round. Scorched by the incandescent heat of the weapon, the world reeled in fever. Elephants were set on fire by the heat and ran to and fro in a frenzy to seek protection from the terrible violence. The water boiled, the animals died, the enemy was mown down and the raging of the blaze made the trees collapse in rows as in a forest fire. The elephants made a fearful trumpeting and sank dead to the ground over a vast area. Horses and war chariots were burnt up and the scene looked like the aftermath of a conflagration. Thousands of chariots were destroyed, and then deep silence descended on the sea. The winds began to blow and the earth grew bright. It was a terrible sight to see. The corpses of the fallen were mutilated by the terrible heat so that they no longer looked like human beings. Never before have we seen such a ghastly weapon and never before have we heard of such a weapon.

Vimanas of Ancient India
The word comes from Sanskrit and seems to be vi-mana = 'apart' or 'having been measured".

A manuscript, written in Sanskrit by King Bhoja deals with different techniques of warfare, and in particular with particular types of war machines. The work is called Samarangana Sutradhara, or "Battlefield Commander (sometimes abbreviated "the Samar"). Chapter 31 is solely devoted to the construction and operation of many different kinds of aircraft with various methods of propulsion.

King Bhoja, who used the Sanskrit term *yantra* to describe the *vimana*, and claims his knowledge was based on Hindu manuscripts which were ancient even in his time, 11th century AD.

Here are some excerpts from this ancient text:

"They were propelled by air and inside is placed a mercury engine with an iron heating apparatus beneath. By means of the power latent in the mercury sets the driving whirlwind in motion. When it has been heated by the controlled fire from the iron containers a thunder power is developed through the mercury. If the iron engine with the properly wielded joints be filled with mercury and fire conducted into the upper part, it develops the power with the roar of the lion."

"Inside it (the Vimana) one must place the mercury-engine with its iron heating apparatus beneath. By means of the power latent in the mercury which sets the driving whirlwind in motion, a man sitting inside may travel a great distance in the sky in a most marvellous manner. "Similarly by using the prescribed processes one can build a Vimana as large as the temple of the God-in-motion.

Four strong mercury containers must be built into the interior structure. "When these have been heated by controlled fire from iron containers, the Vimana develops thunder-power through the mercury. And at once it becomes a pearl in the sky.

"Moreover, if this iron engine with properly welded joints is filled with mercury, and the fire be conducted to the upper part it develops power with the roar of a lion."

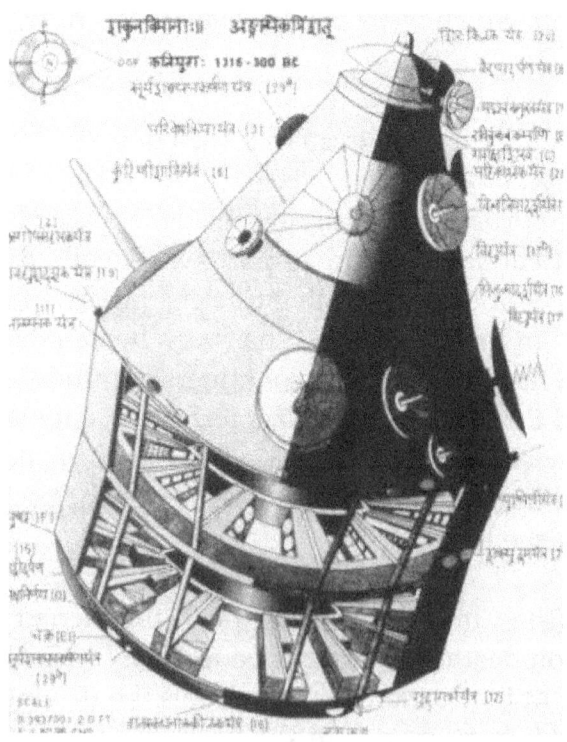

Taltala: Maya Danava was described as having given a vimana to the Indian King Salva and was a being from the planetary system called Taltala Although I could not find any planetary or star system having that name; Taltala is commonly used in India and was given as the name of the sea god of Atlantis, also known as Poseidon.

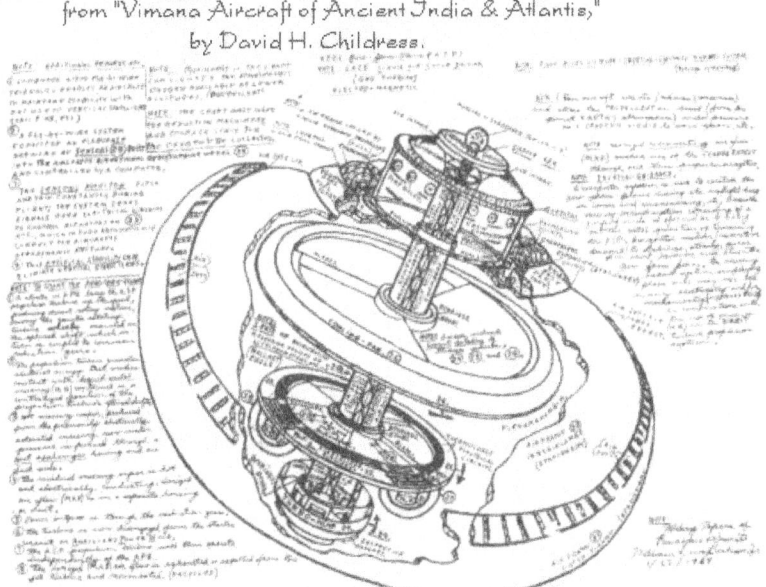

from "Vimana Aircraft of Ancient India & Atlantis," by David H. Childress.

W.D. Clendenon's illustration of his concept of a Mercury Vortex Discoid craft. This is his basic concept of the propulsion of at least one type of Vimana. Many current UFOs may well use a variant on this design. Clendenon believes that George Adamski's craft was also a Mercury Vortex Vimana.

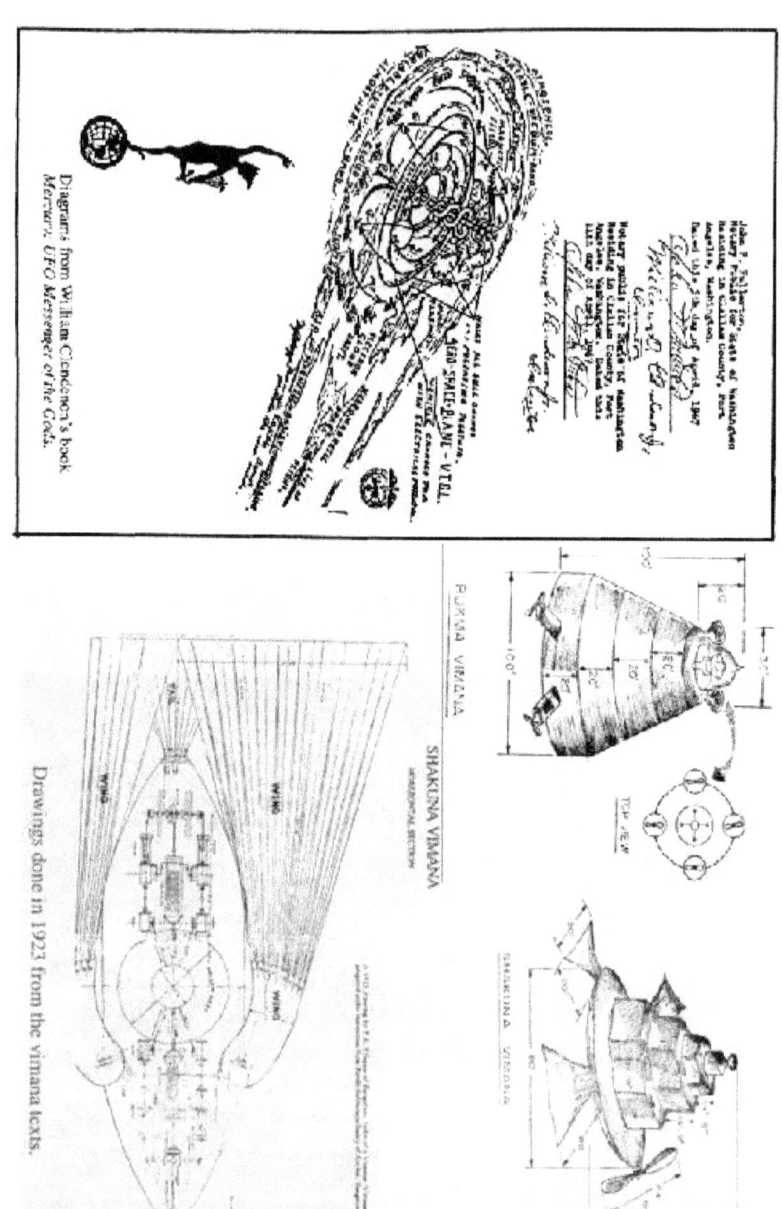

Diagrams from William Clendenon's book *Mercury, UFO Messenger of the Gods*.

Drawings done in 1923 from the vimana texts.

ELECTRIC POWER GENERATOR
TOP VIEW

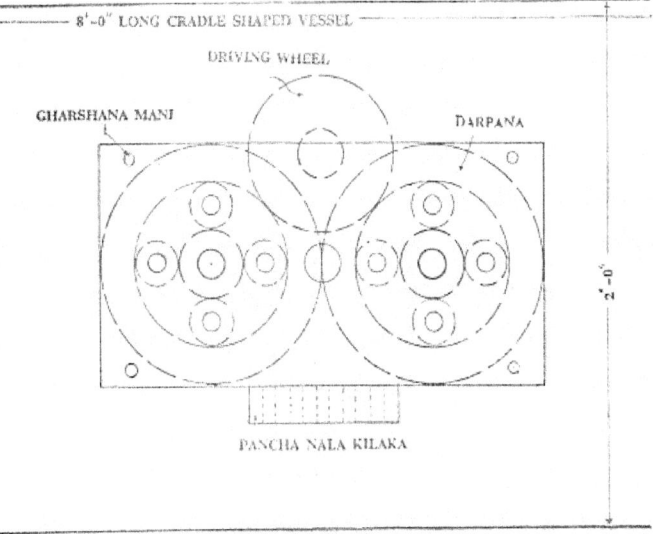

ELECTRIC MOTOR

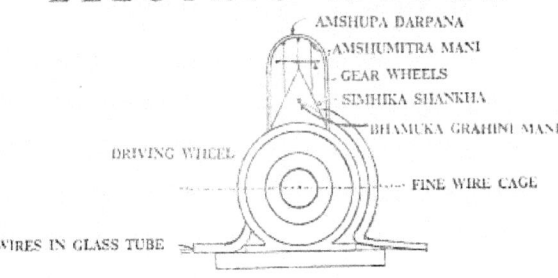

ELEVATION

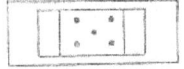

PLAN

Drawn by
T. K. ELLAPPA,
Bangalore.
2-12-1923.

Prepared under instruction of
Pandit SUBBARAYA SASTRY,
of Anekal, Bangalore

ELECTRIC POWER GENERATOR

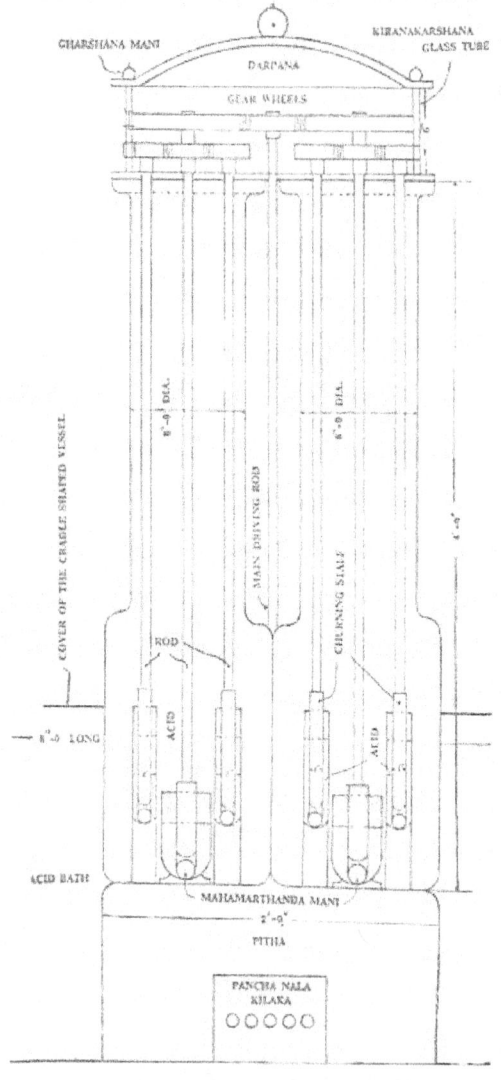

SECTIONAL ELEVATION

Drawn by
T. K. ELLAPPA,
Bangalore.
2-12-1923.

Prepared under instruction of
Pandit SUBBARAYA SASTRY,
of Anekal, Bangalore

Please watch this video so you can get an insight into what I am trying to explain to you. More information is on the Ancient Astronauts and in the Resource section.

http://www.themeasuringsystemofthegods.com/THE%20VYMAANIKA-SHAASTRA-vinama%20device.flv

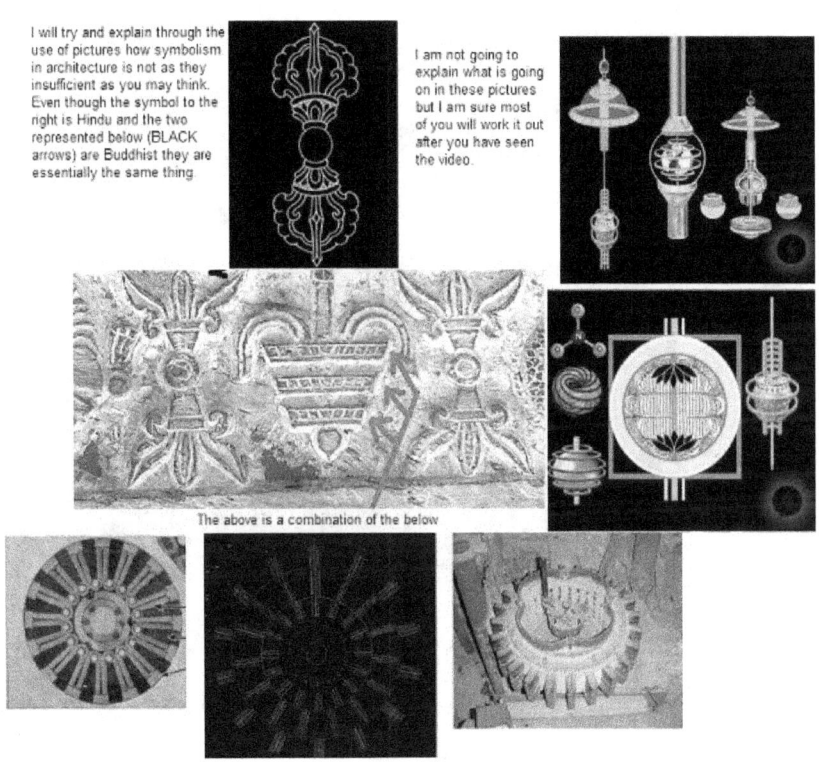

Hitler and the Thule and Vril Societies
Look at the man behind Hitler with the pointy hat on. The Thule and Vril Societies used Occult knowledge to obtain this technology from Tibet and India.
http://discaircraft.greyfalcon.us/

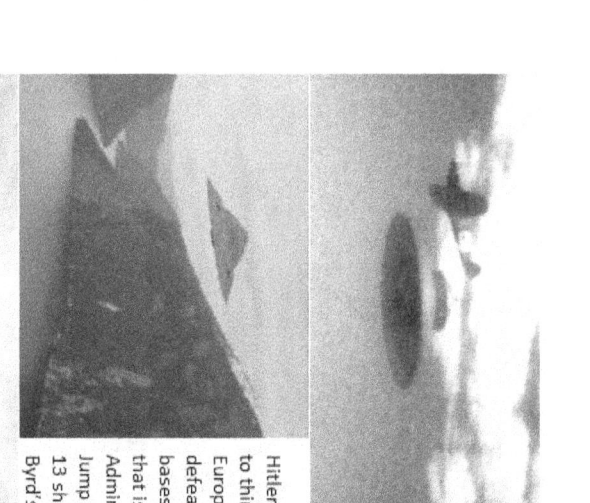

These craft where far superior compared to anything English or the U.S had.

Hitler Spread his force out to thinly and lost the war in Europe but was never truly defeated. The Germans had bases down in Antarctica and that is where they defeated Admiral Byrd in Operation High Jump this included 4,700 men, 13 ships., and multiple aircraft. Byrd's expedition ended after 8 weeks with "many fatalities" according to initial reports based on interviews with his crew. Admiral Byrd revealed in a press interview that Task Force 68 had encountered a new enemy that "could fly from pole to pole at incredible speeds."

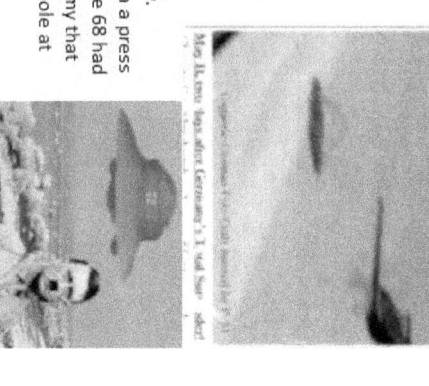

There are many Videos on YouTube about this but the best one is in Russian (no English)
Also look for Admiral Byrd's Diary to find out what else happened down there.

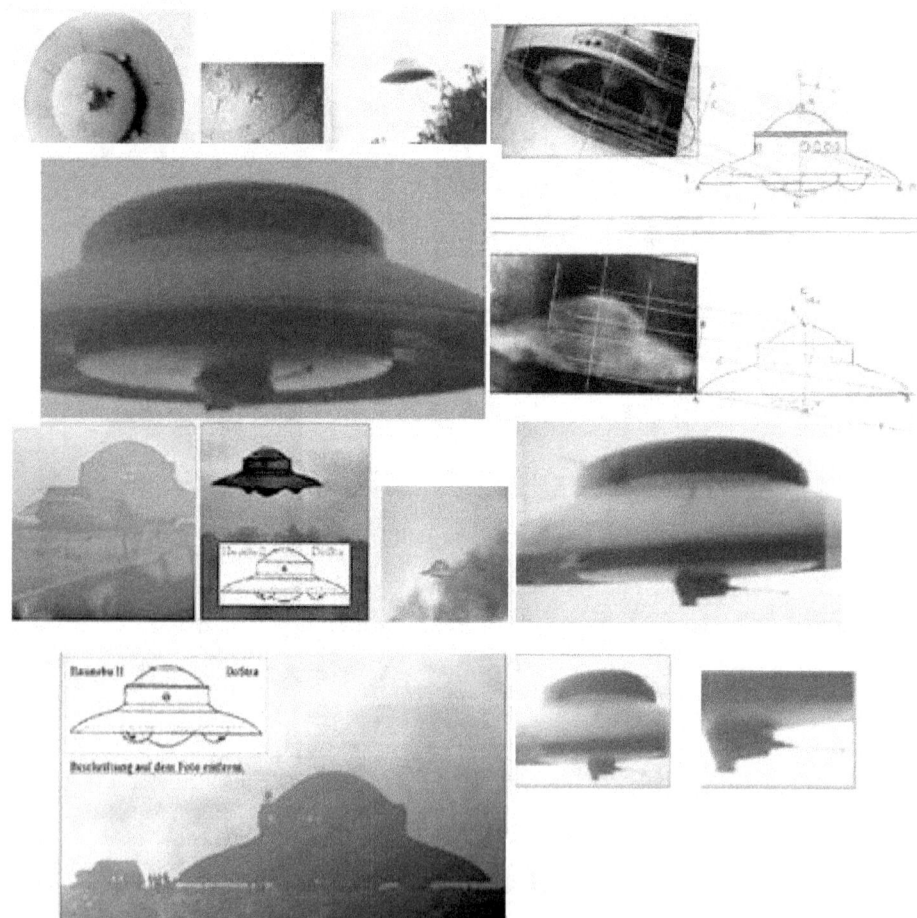

Now I am going to give you something to think about. There are three possibilities here for you to consider and I have tried to provide adequate information for the reader to form their own opinion. I know this is not a popular subject but the agreements are very thought provoking. One important thing to remember is Maya is the illusion of a limited, purely physical and mental reality in which our everyday consciousness is fooled by the limited range of our senses. Knowing that your eyes only see information (code) and our mind interprets this into images to manifest a three dimensional world within our mind.

Flat Earth and Gravity

In the last 20 years Quantum Physics (pseudo physics) has

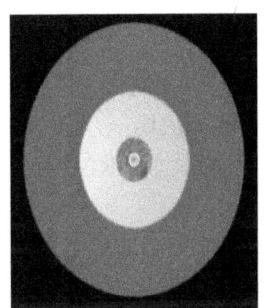

been reverting to a two dimensional model of gravity to understand higher dimensional planes of existence. Most philosophers are aware of higher planes of consciousness and being planes imply that they are flat like a disk as Vedic philosophy shows.

Vedic philosophy also states the wold to be a Globe, which would be a 3-D projection of a 2-D plane of existence. Images are from the video "Vedic Tour of our Universe and Beyond" according to the "Srimad Bhagavatam" As you can see on top of the mountain there is a square that represents the earth.

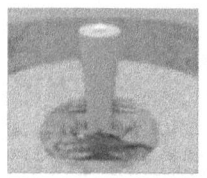
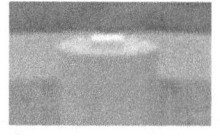
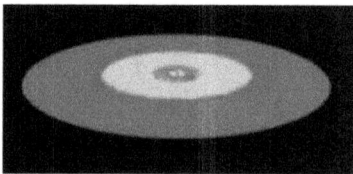

Below is the Vedic 3-D world and then we are continuing on to the Christian interpretation of Vedic philosophy.

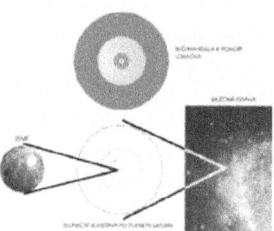

Even in the Bible the four corners of the earth is mentioned and the ends of the earth.

Isaiah 11:12 and he shall set up an ensign for the nations, and shall assemble the outcasts of Israel, and gather together the dispersed of Judah from the FOUR CORNERS OF THE EARTH. (KJV)

Revelation 7:1 And after these things I saw four angels standing on FOUR CORNERS OF THE EARTH, holding the four winds of the earth, that the wind should not blow on the earth, nor on the sea, nor on any tree. (KJV)

Job 38:13 that it might take hold of the ENDS OF THE EARTH, that the wicked might be shaken out of it? (KJV)

Psalm 104:5 "He set the earth on its foundations; it can never be moved." (Fixed in one place)

Ecclesiastes 1:5 "The sun rises and the sun sets and hurries back to where it rises."

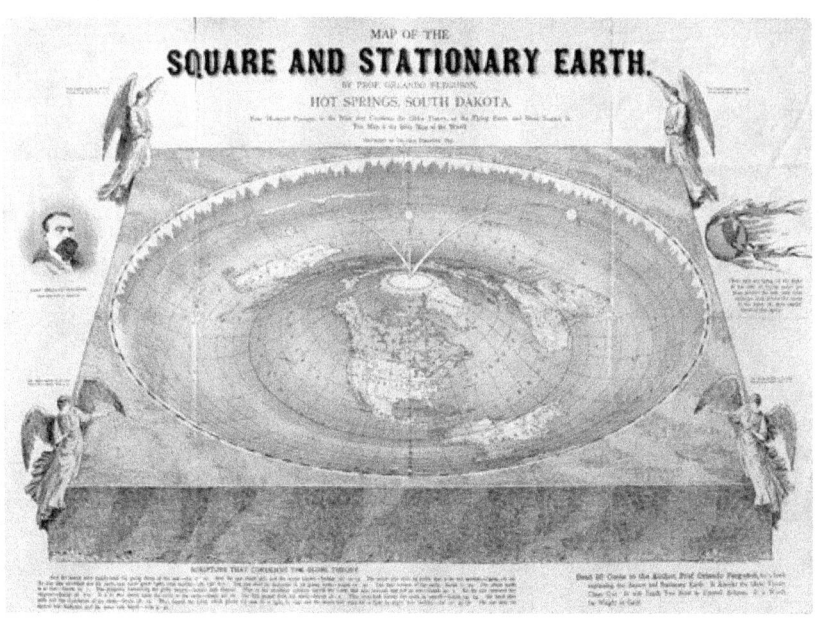

Now keep in mind the four corners and edges of the earth and the sun hurrying back to where it rises when looking at the following image.

Map of the square and stationary earth. By Orlando Ferguson, 1893.

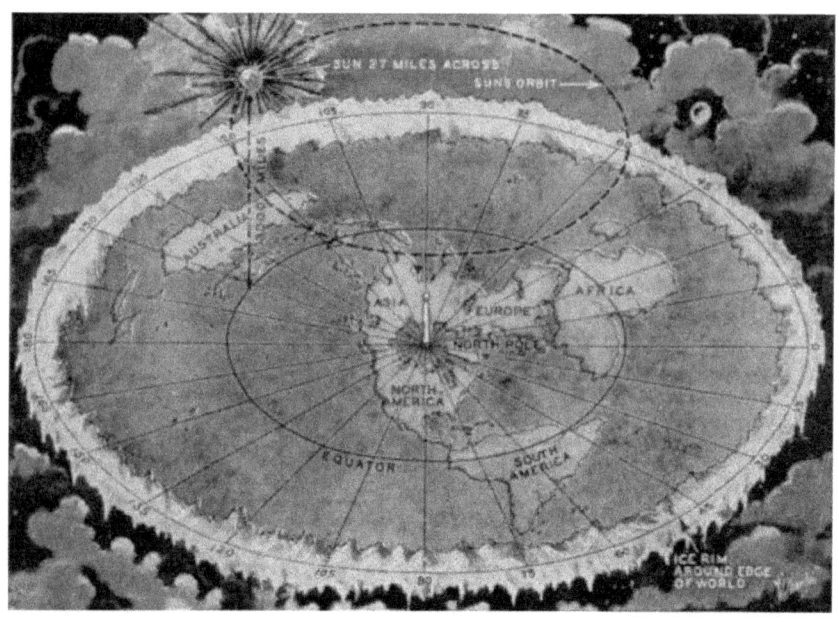

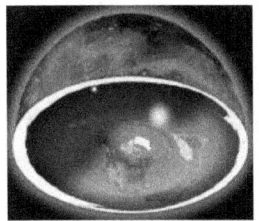 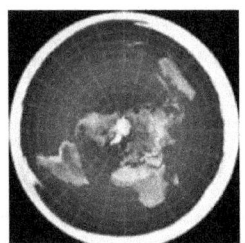 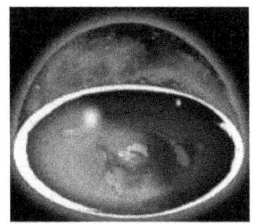

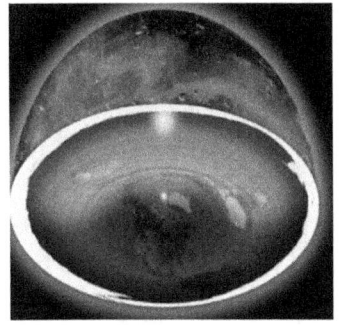

The flat earth with the sun and moon revolving around the earth as it should be. Antarctica is basically closed and off limits unless you apply for permission to go there, even commercial planes don't fly over it.
It has been stated that it is impossible to cross Antarctica, if it were a continent as the pseudo scientists say it is there is no reason why it can't be crossed. Antarctica is the boundary of the flat earth.

http://theflatearthsociety.org/cms/

The founders of the United Nations use the flat earth map not a globe.

THE FLAT EARTH SOCIETY

President:
W. Mills,
7 Vale Grove,
Finsbury Park, N.4.

Organizing Secretary:
S. Shenton,
22 London Road,
Dover.

o o o o o o

The International Flat Earth Society has been established to prove *by sound reasoning and factual evidence* that the present accepted theory, that the Earth is a globe spinning on its axis every 24 hours and at the same time describing an orbit round the Sun at a speed of 66,000 m.p.h., is contrary to all experience and to sound commonsense.

In ancient times the Earth was regarded as plane, and this is expressed in all literature up to a few hundreds of years ago. The theory has fallen into disfavour, owing mainly to the dogmatism of modern science and popular education in schools, which leads to prejudice in favour of the globular theory from the start.

It is always a pity to allow false theories to pass unchallenged, and it is hoped that the Flat Earth Society will do much to undo the harm that has been caused. Remember that the truth of the plane figure of the Earth can be shown *by irrefutable evidence,* and anyone who is interested in becoming a member is asked to contact the President or the Organising Secretary! In future, it is hoped to hold regular meetings of the Society.

December 20th, 1956.

Modern pseudo-Science is full of cover-ups, lies and just plain fraud. People need to investigate everything they are taught and remember that science is based on observations made by the senses. These are only electrical impulses sent to the brain not what is really going on within the illusion.

Arne Sivertsen. Lecturer at a Teachers' Training School, Norway. Apl:67.
" I have never really believed what I have to teach my pupils".

Steven Karwash and Stephen Feld. New Castle, Pa. USA. March 67.
Re. American reports on Society's arguments concerning earth's structure. "These contentions are quite thought-provoking and cause one to seriously consider whether or not we have been subjected to gross misinformation from our respective governments." Similar doubts have been many times expressed by thoughtful people.

W.L.C. NASA Goddard Space Flight Centre. USA. Oct: 66.
" It occurs to me that the feelings which you conveyed in your statement are not peculiar to yourself and those in your organisation, BUT ARE IN FACT QUITE UNIVERSALLY FELT, IF SELDOM EXPRESSED."

J.H.M. Lead engineer, Radiation Corptn; Florida. USA. Sept 66.
"... engaged in Space work, I feel that I will be able to materially further the work of the Society from within the enemy's lines, as it were."
" There is much to be done!"

J.R. Missile & Space Division, Penna. USA. Sept. 66.
"It would be helpful to know your theories TO HELP SUBDUE THE PANIC IN THE EVENT THEY SHOULD BE REVEALED AS FACTS!"

Instructor and full physics class, Eldora, Iowa. USA. Feb. 67.
(And many similar classes throughout USA and Canada.) "The nature of our concern lies with the un-scientific explanations of earth's spherical shape".

Dave Sannes, from Mortar Bunker, Vietnam. May 67. "Faith. - Belief without evidence, in what is told by one without knowledge, of things without parallel". (Ambrose Bierce.) . Mr Sannes continues:- "Such was the nature of my innocent faith that the world was round"! Our thoughts are much concerned with this young man in what he calls,"this combat theater".

Stefan Pomoak, Czechoslovakian Scientist. May 67. mentions:-
"...the actual position and extent of our earth".

N.B. If corresponding, please enclose International Postal Reply coupons.

The reason why we are fool by our senses is because what N.A.S.A scientists call the EM field around the earth is an Artificial intelligence net creating this illusion which was known as the 12 Aether's from the ancient Greeks. There is no light in space the stars are can only been seen from under the earth's atmosphere and the sun is a converter, a transformer of sorts.

Scientific Blasphemy

The following is proof, for those innocent minded Christians who still need it, that " Science " has something to do with the question of Salvation, inasmuch as it is leading men not only to deny the truths of the Bible, but, as a consequence, to deny the Christ and the God of the Bible. The first paragraph is from a weekly paper with the very suggestive title, Lucifer, published in America; and from a number dated "December 23, E.M., 287." This is instead of calling it the year A.D. 1887,
Why do they refuse to acknowledge the A.D.? The editor himself shall tell us.

He says:—
"We date from the First of Jan. 1601. This era is called the Era of Man, (E. M.) to distinguish it from the theological epoch that preceded it.

In that epoch the earth was supposed to be flat, the sun was its attendant light revolving about it.

Above was Heaven where God ruled supreme over all potentates and powers, below was the kingdom of the devil, hell. So taught the Bible. Then came the new astronomy. It demonstrated (?) that the earth is a globe revolving; about the sun; that the stars are worlds and suns; that there is no "up" and "down" in space. Vanished the old heaven, vanished the old hell; the earth became the home of man. And when the modern cosmogony came, the Bible and the church, as infallible oracles, had to go, for they had taught that regarding the universe which was now shown (supposed?) to be untrue in every particular." Gently, friend Lucifer, for you is somewhat in the dark here, notwithstanding you assume to be a great light bringer! It never has been "demonstrated" or "shewn" that the earth is a whirling globe, and that, therefore, the Bible cosmogony is wrong. It has been quietly assumed by the "new astronomy" and the assumption has been cowardly yielded by the "Christian" who ought to have challenged it. But it never has been proved.
Never! I f it has, kindly give us the name and the address of the man who "demonstrated" it. Newton and Copernicus, both, were candid enough to confess that the theory called by their names is but a theory, a mere assumption not based on known facts. Their disciples forget this.
The Sun Standing Still, AT the COMMAND OF JOSHUA, (Albert Smith)

CIRCUMNAVIGATION

Many people foolishly imagine that ships can sail in a straight line due E. or W, but if a line be drawn, all round a sphere; it would make a circle, a chalk mark round a football for instance. A circle is not a straight line, as I once had reason to remind an educated gentleman in a public debate. He was known, too, as "the Leicester astronomer!" In the above figure the magnetic north "pole" is represented at N; and if a ship, sailing round the outer circumference, keeps the point of the compass always towards N, and steers at right-angles to it the course described will be a circle. A small flat island could be circumnavigated in the same way, with a powerful magnet in the middle of the island; the ship thus describing a circle. But if a vessel took a straight line course from A, it would sail in the south-westerly direction towards S.W.

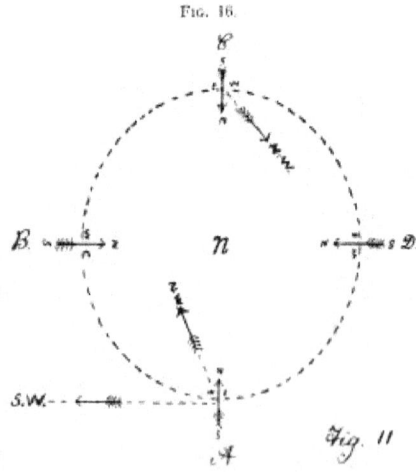

On a globe it would be impossible for the horizontal needle always to point to the north magnetic "pole" from different parts of a spherical sea, as anyone may prove by laying a needle at various points as a tangent to a large ball. But on a flat surface the needle always points to the centre while the ship describes a circle— which double fact not only again explodes the globular theory, but establishes the truth o f a plane earth and sea! We have years ago many times pointed out this fact in our literature, and as a result one professor has had the honesty to make the following confession;—

"The earth has been circumnavigated a great many times. ... We can (we could?) journey round the globe, sometimes travelling on land, and sometimes on the sea. This would appear to be a certain proof that the earth's surface is curved. Nevertheless it has been pointed out that circumnavigation would be possible if the earth had A FLAT SURFACE with the north magnetic pole at its centre. A compass needle then, would always point to the centre of the surface, and so a ship might sail due east and west, as indicated by the compass, and eventually return to the same point by describing a circle." (Caps, mine.)— Prof. R. A. Gregory, F.R.A.S., *Elementary Physiography*'.

Yet thoughtless teachers still refer to the schoolboy proof that circumnavigation proves the earth a globe!

The least intelligent of our *readers will* hardly need much explanation *to* understand *the chart* an amateur friend has kindly sketched out for us. Take it to any part of the world, the mariner's compass uniformly points to the central North. Navigators have announced the strange fact that inside the frozen belt of Northern icebergs', varying from 80 or 90 to 100 miles in breadth, is an unfrozen sea, upon whose bosom no craft of man in any shape has ever rested, the distance over the ice Rendering the transport of any vessel physically impossible.

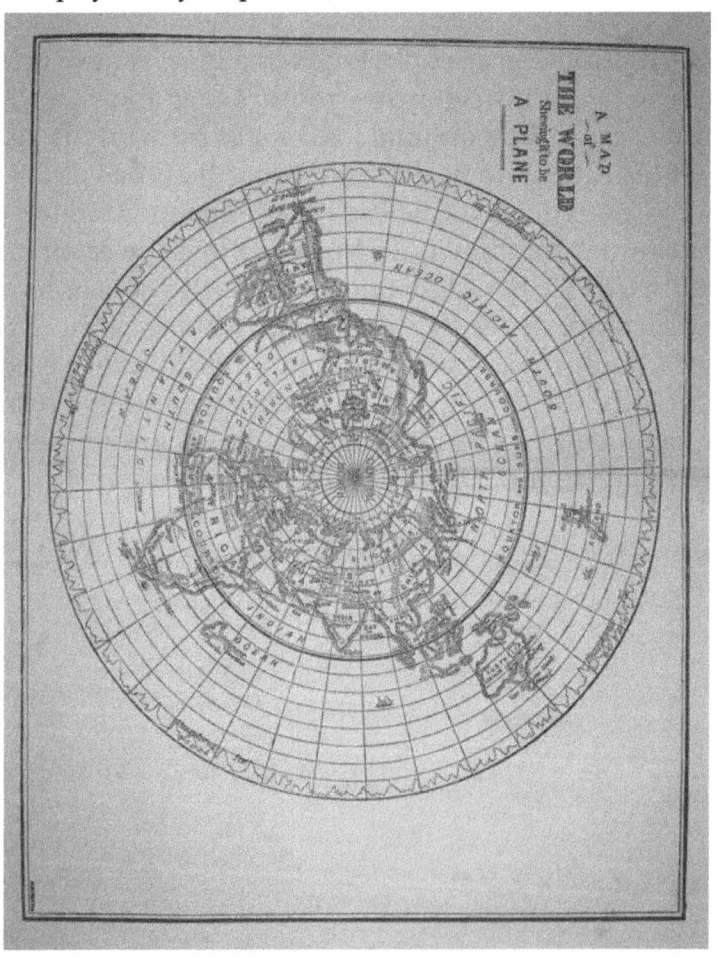

The cause of its being unfrozen has never been ascertained, and no surmise can be offered beyond the supposition that a submarine volcano or hot springs must cause a higher temperature of the waters at that point. The well-known length of the day alternating between the Northern centre and the Southern circumference is caused by the contraction and expansion of the Sun's orbit—nearer to the North in summer, and more distant in the winter months. This will be understood more clearly on reference to the larger work, where phenomena of day and night, summer and winter are fully explained.

For those who have "eyes to see"
ZETETIC COSMOGONY
REVEALS :-

The "PLANE" TRUTH

A modest pamphlet with a great aim.
To re-establish in young people, FAITH in The Creator and a truer conception of their earth environment.
• • • • • • • •

The International Flat Earth Research Society, DOVER. England.
With modern events and observations added to the basic work of the former "Universal Zetetic Society" of U.S.A.& England.

Passing over the Equator, we come to the frozen extremities of the world—South, South, South all the way round. Pacing the Northern centre from any point of the Southern circumference, of course, to the right is East, and to the left is West. By sailing *due* East or due West, the ship returns again from an opposite direction to that in which it set out on its voyage. *But* it can neither pass the Northern, much less the Southern barriers of icebergs. What lies beyond the outer circle, no *man has* yet dared to explore; and but a few miles into the airy regions of space may the bold aeronaut ascend; so into the icy barriers below, the most daring adventurer is told—" Thus far shalt thou go and no further." Further knowledge would be too much for the finite mind of man to bear. His all-wise Creator, in pity to his frame, has given him limits, which he cannot, dare not over-pass. "Such knowledge is too wonderful and excellent for him; he cannot attain to it."

Let him be content with the wonders his Lord has seen fit to reveal. But the .space above him, beneath him, and all around, is, in this stage of his existence, hidden from his view.

The Popularity of Error and the Unpopularity of Truth. BY JOHN HAMPDEN , ESQ 1869
The Plane Truth, by Samuel Shenton 1966

Remember that all reproductions from photographs taken from rockets or capsules are as the camera recorded the events, and NOT as viewed by the astronaut's human vision. The camera distorted horizons have always been a misleading factor with those who have not freed their minds from the "Planet" or Globe earth indoctrination.

Three or four years ago, the U.S.I.S. booklet "Science Horizons", the same information service booklet as the one which published the photomosaic/
Carried a note to the effect that the Americans hoped to produce a lens which would NOT distort level horizons. So far I am not aware that such aid to truer photography has yet been made available. Flat Earthiest however can prove that due to the known laws of perspective, the horizon, optionally, rises and remains level with the observer's, or the camera's, eye, no matter what height is achieved. In foot the earth immediately beneath balloon, airplane, rocket or capsule, presents a dish-shaped or concave appearance. The point of earth immediately below the vehicle is the lowest.

It is NOT the highest point of your "globe" earth with the dip or curvature of the "ball" sweeping away downwards to a horizon far away below the eye level.

NASA's symbol shows their forked tongue.

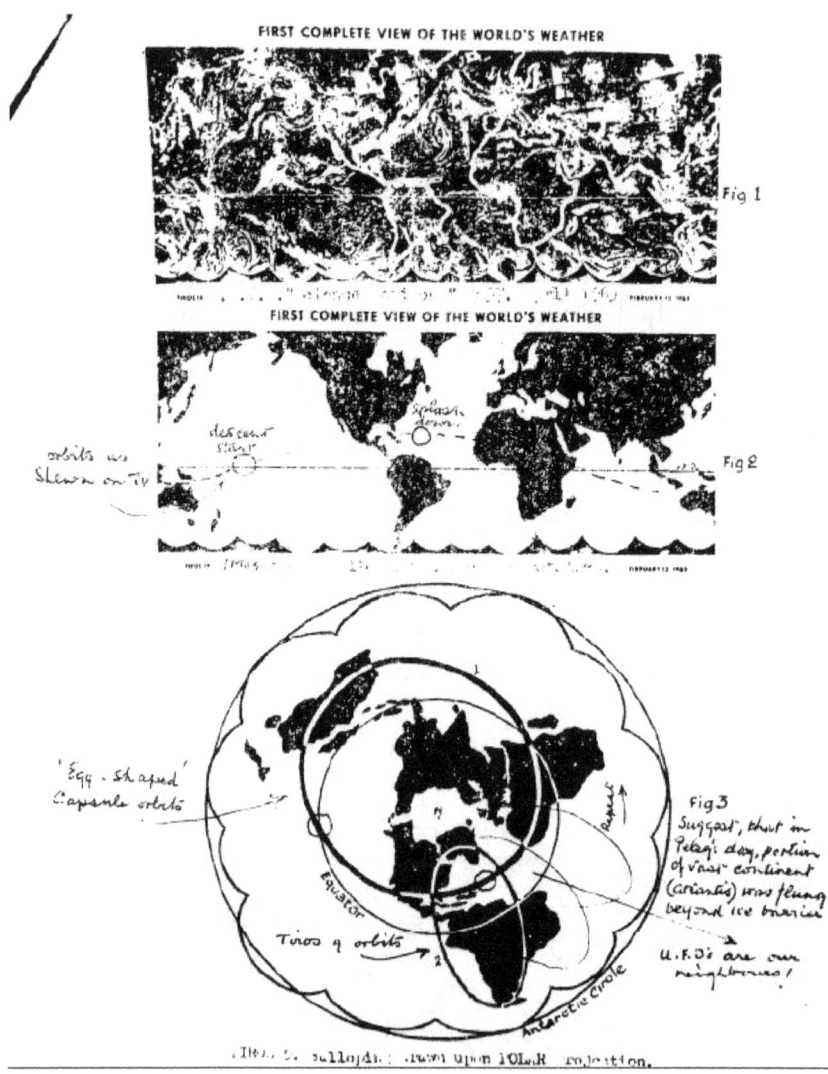

Article from Modern Mechanics, year unknown. Voliva's conception of a flat world, with the North Pole in the centre and the sun revolving in its orbit above the equator. A wall of ice around the edge of the earth keeps adventurous mariners from falling off into space. **By JAY EARLE MILLER**

$5,000 for Proving

Wilbur Glenn Voliva, who claims the earth is flat and offers $5,000 for proof to the contrary.

If you can prove that the world is a sphere, floating in space, turning on its own axis, revolving around the sun, you can earn a prize of that amount Such a prize has been posted for years, offered by Wilbur Glenn Voliva, general overseer of Zion, lll, home of the Christian Catholic Apostolic Church, founded some thirty years ago by the late John Alexander Dowie. Post and Gatty did not fly around the world, according to Wilbur Glenn Voliva; they merely flew in a circle around the North Pole. This article presents Voliva's theory of a flat world, and tells you how you can win his offer of $5,000 for proving that he is wrong.

Many have tried to claim the $5,000 — and all have failed. The catch is that your proof must not start with the assumption that the world is round, or rather a globe, for Voliva believers the world is round, but a round, flat disc rather than a sphere.

the EARTH a GLOBE

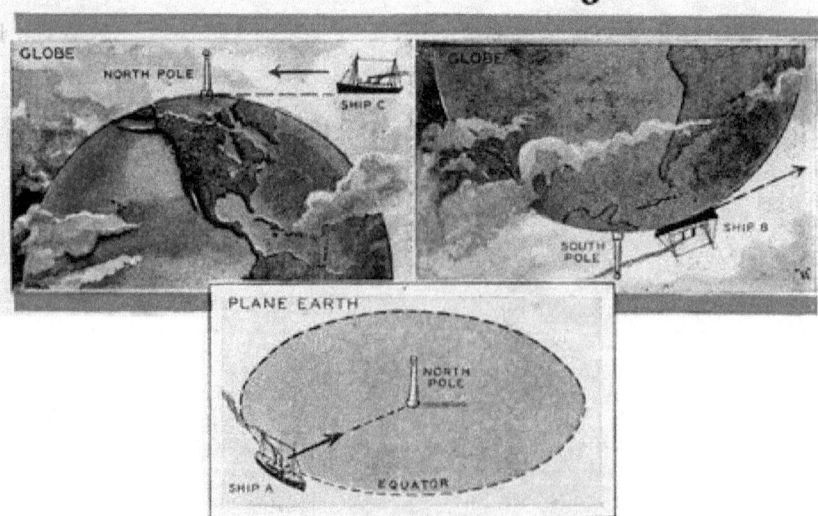

Without the basic premise that the earth is spherical, no one has found an absolutely convincing proof that Voliva's wrong when he describes his disk shaped world, firmly planted on its foundations, surrounded by a wall of ice to keep mariners from falling off the edge, and surmounted by a crystal dome in which the stars are hung like chandelier to light the night.

These three drawings are what Voliva offers as graphic proof that the earth is not a globe. Ship C above represents an impossible position in space which must be occupied by any ship that is hundreds of miles from the pole, and still has the compass needle pointing north from a level position. What is your answer? Ship A is represented at about 45 degrees south latitude, and show, how the compass needle can point north without distortion on a fiat world. Ship B is represented above in the southern hemisphere, and shows how impossible it is for the compass needle on this ship to point north, for a compass never indicates north to he down in the ground. This is an argument you'll have to answer convincingly if you win the $5000 offered by Voliva.

Nor can you submit proof to absolutely disprove the belief of Voliva that the sun, instead of being an 800,000-mile ball of fire more than ninety millions of miles away is really a insignificant affair, only some 27 to 30 miles in diameter and about 3,000 miles above the earth. Or that the sun and moon move in orbits while the earth stands still, that the moon is about the same size as the sun and the same distance from the earth, shines by its own light, and moves in much the same orbit as the sun.

In these days of 100 and 200-inch telescopes accurate measurement of the speed light and the diameter of distant stars, any challenge of the correctness of the familiar Pythagorean -Copernican- Newtonian systems of astronomy may come as a shock. Actually, the belief that the world is flat is not confirmed to Zion. China, which had workable systems of astronomy as soon if not sooner than Greece or Egypt, continues to stick to the flat world theory, and thousands of other people scattered over the globe have this, or theories even more unusual.

Ever hear of Orlando Ferguson? Fergus on, a resident of lint Springs, South Dakota designed a weird and wonderful map back in 1893 which showed a square world, with an angel seated on each of the four corners. The inhabitable or known world was not, however, flat. The central section, comprising what we know as the northern hemisphere, was convex, curving downward from the central "North Pole" to the equator, while the southern hemisphere was concave, curving upward from the equator to the rim. The shape was much like a soup plate with a raised centre.

Then there is Gustave F. Fbding of Cleveland, who recently published a book to prove that the world is a hollow sphere and that we live inside it instead of on the outside, the so-called Koreshan theory of Koresh and Prof U. G. Marrow. Then there is the theory of Marshall B. Gardner, that the earth is sphere with the inhabitants living on the outside, but that the sphere is hollow. Variations of the Gardner theory have been used often by fiction writers, who have peopled the inside with strange races and hung up a small sun in the centre to illuminate it.

Aristotle's Proofs That the Earth Is a Globe and Voliva's Refutations

The changing aspect of the heaven in different latitudes, some stars appearing and other disappearing proves the earth is round. Or does it?

"The stars are set in a hemisphere dome so close the earth that all cannot be seen at the same time," is Voliva's explanation of this point. Compared to some of ideas Zetetic Astronomy or the flat world theory of Voliva and his followers at Zion has considerable to commend it. Aristotle in ancient Greece cited the three commonest proofs that the earth is a sphere 1800 years before Columbus "sold" the world on the spherical theory, and they are still use. The mass of the people did not really begin to believe that their world was a globe until the earth had been circumnavigated.

Actually, that was out as good a proof as any of the three cited by Aristotle, for it is possible to circumnavigate Voliva's flat world, travelling either east or west, and come back to the starting point.

Aristotle's three points were: **First:** The disappearance of ship as it sales over the horizon, the hull vanishing first, and the mast and rigging last, and the reappearance of an approaching ship in the reverse order serve order. **Second:** The curving shape of the earth's shadow on the moon during an eclipse. **Third:** The changing aspect of the heavens in different latitudes, some stars disappear and other appearing, as the polar star in the northern hemisphere and the Southern Cross south of the equator.

"There is not a scintilla of truth in any of them," Voliva retorts, 'and yet 'you will find then in every geography, and every primary teacher repeats them like a parrot. I decline to be a parrot. A parrot is a man who never thinks for himself, but repeats what he hears without any questions as to why or wherefore." Zion maintains that the disappearance of a ship over a horizon hull first is an optical Illusion of perspective, no different from the apparent merging of the railroad tracks in the distance. A man at the foot of a tree a couple of miles across a plain may be invisible, while the tree itself stands up against the sky and is visible for miles.

Earth curvature of eight inches to the mile is not sufficient to explain the invisibility of the man. As for Aristotle' second point, Voliva and his followers maintain there is no proof that the curving shadow of an eclipse is the shadow of the earth, and maintain that there have been several eclipses within historical times in which both sun and moon wore visible at the same time, so that the eclipse could not have been due to the earth's shadow.

As for the third proof, the Zionites and other believers in the Zetetic Astronomy of "Parallax" maintain that the stars are set in a hemispherical dome so low and close to the earth that not all stars can he visible from any one point. Dr Samuel B. Rowbottom, of England, who, under the name of "Parallax" provided the examinations of all natural phenomena to fit the flat world theory, died in 1884, but his followers have keep his work alive.

Voliva and Scientists Prove Different Theories From Same Phenomena

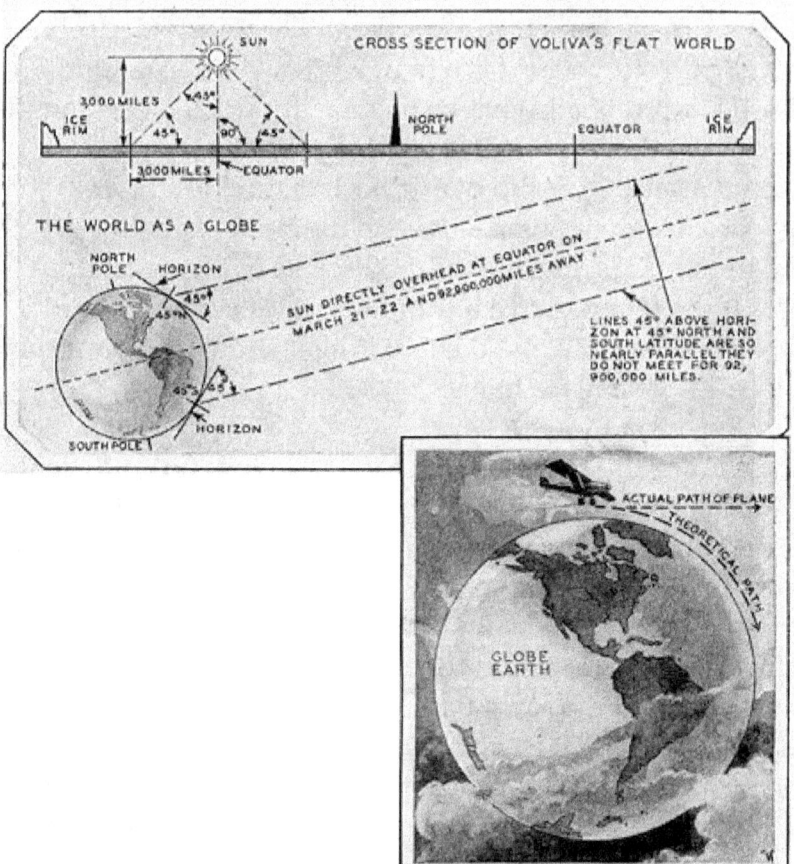

The spring and autumn equinox as seen from 45° North and South Latitude or, either a flat or a spherical earth. Voliva's world is a disc 24,000 miles in diameter. On March 21-22, the sun is directly above the equator and is seen at 45 above the horizon at 45 degrees North and South Latitude. The distance from the equator to 45 degrees either North or South Latitude is one eight of the earth's diameter, or 3000 miles; therefore, the sun must be 3000 miles away.

With the spherical world the same reasoning would place the sun 92,900,000 miles away. The diagram explains both theories.

'An airplane is built to fly level and follow a horizontal line. When tying 300 miles an hour on a globe an airplane pilot must drop his ship 60,000 foot an hour from a level line drawn tangent to the earth at the point of departure."

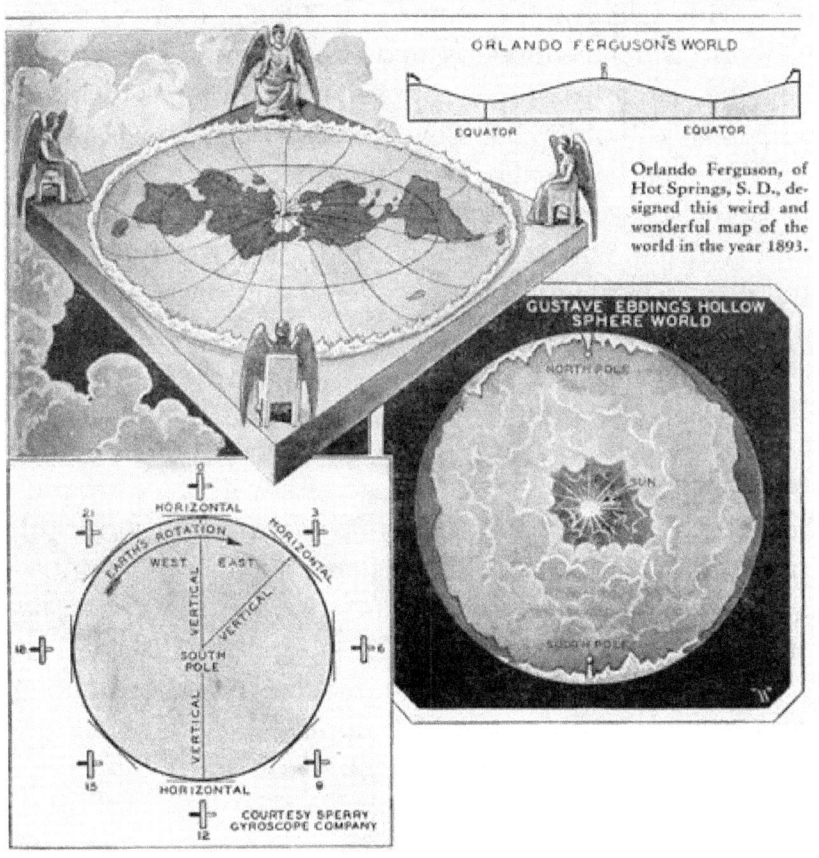

Voliva Is Not the Only Man to Have Weird Ideas About Earth's Shape

Orlando Ferguson, of Hot Springs, S. D., designed this weird and wonderful map of the world in the year 1893.

A gyroscope at the equator with its spinning axis horizontal in an east and west direction, will appear to make one evolution a day about an axis at right angles to the spinning axis of the earth.

Gustave F. Fbding of Cleveland, who recently published a book to prove that the world is a hollow sphere.

The flat world theory in not confined to any country, sect or group. Within recent years, the Rev. John Dmich A Catholic priest has written a book "the earth is not round" which has had a wide sale. Another pamphlet, "One Hundred Proofs the world is not a Globe", issued by William Carpenter in 1885, the year after Rowbottom's death, continues to appear in revised editions. A book by Alexander Gleason, issued in 1893, is a standard text book among the Zetetic believers. Gleason also issued the map that is used in the parochial school at Zion, where the flat world theory in taught as fact and the spherical world belief covered only as an incidental superstition.

The Gleason map, a copy of J.S Christopher projection shows the world as disc, with the North Pole in the centre and the South Polar Region, spread as a wall of ice around the rim. The northern hemisphere is more nearly correct in this form than on the standard Mercator project generally used in schools, for navigation and for other purposes. The southern hemisphere, however, is enormously distorted. Australia, for example, is drawn out of an enormously long and narrow island. Chester M. Shippey, Director of research in Voliva's church cabinet, this is one of the weaknesses of the flat world theory, for the Time table of the Australian transcontinental railroad shows a mileage far less than the Christopher map would call for.

Voliva maintains that there is no South pole, and that ii is 60000 miles around the southern ice wall, Captain Gunnar lsachsen, the Norwegian explorer, last winter circumnavigated navigated the Antarctic continent in a voyage if about 14000 miles. Zion says lsachsen, may have circumnavigated something possibly an island of that size, but did not go around the Antarctic ice rim, and points to the 60000 mile journey of Ross in 1848 and the following two years, when he circumnavigated the ice rim. Of course floss was in a sailing ship which tacking back and forth for three years on a journey Isachsen completed in few weeks, which could explain the discrepancy in the distance travelled.
"They say that Admiral Byrd flew over the South Pole,"

Voliva said recently, "But there is no South Pole. East, West and south are not absolute directions; they are only relative directions with reference to North. East and West are points at right angles to north."

Voliva maintains that the sun is no more than 30000 miles in diameter and about 3000 miles from the earth. As proof, he points to the fact that on March 21-22 the sun is directly overhead at the equator and appears 45 degrees above the horizon at 45 degrees north and south latitude. As the angle of sun above the earth at the equator is 90 degrees while it is 45 degrees at 45 degrees north or south latitude, it follows that the angle at the sun between the vertical from the horizon and the line from the observers at 45 degrees north and south must also be 45 degrees.
The result is two right-angled triangles with legs of equal length.

The distance between the equator and the points at 45 degrees north or south is approximately 3,000 miles, so the sun would be an equal distance above the equator and, from the apparent size of the sun's image, it would follow that it has a diameter of about 30 miles.

Of course, if one starts with the assumption that the world is a sphere instead of a flat surface, the same facts can be used to prove that the sun is nearly 93,000,000 miles away and has a diameter of more than 800,000 miles

Voliva argues that if the sun were that big and at that distance there would be no change of seasons because the sun's rays would reach both hemispheres with equal volume regardless of its position north or south in relation to the equator. On the other hand the same argument can be used to prove that on his flat world there could be season, and no period of total darkness at the North Pole. Actually, of course, the amount of the sun's rays reaching the earth is radically reduced by the envelope of air arid moisture, without which, in fact, life would be impossible, because days would be unbearably hot and nights impossibly cold, just as they are on the moon and some of the planets. Prof Piccard, the German balloonist, found a temperature of about 72 below zero outside his airtight balloon car at an altitude of ten miles, while inside it the temperature was 104 degrees above zero.

If the flat world believers deny the absorption factor of moisture and air in the round world theory they have removed the only argument that could explain seasonal changes and polar nights in their flat world.

Some of the claimants of Voliva's $5,000 have argued that, because the moon and the planets appear to be spheres it must follow that the earth is a sphere, an assumption that Voliva and his followers deny. That says Voliva, is like arguing that because a COW is an animal and has horns, all animals have horns, or that all COWS have horns.

Probably the best spherical world proofs ever found were the two discovered by Jean Bernard Leon Foucault, the famous French engineer, when he invented the Foucault pendulum and the gyroscope. The performance of both can only he explained on the assumption that the earth is a sphere, revolving on its axis, but they do not prove the fact within the meaning of Voliva's prize offer. The Foucault pendulum illustrates the diurnal motion of the earth as it revolves on its axis, the plane of the oscillation of the freely suspended pendulum slowly changing until it appears to make one revolution each day.

The gyroscope has the same property. If a gyroscope is set spinning on the equator with its spinning axis horizontal in an east and west direction it will appear to make one revolution a day about an axis at right angles to the spinning axis. At the end of twelve hours the gyroscope will appear to have reversed ends, though actually it will continue to point just as it did at the start, only the earth will have made a half revolution.

Also, if a spinning gyroscope were carried around the earth along a north and south meridian, passing over the two poles, it would constantly change its angle so that the horizontal spinning axis would always be at right angles to a vertical line from the earth's centre. In other words, if the gyroscope were at 50 degrees north latitude, the gyro axle would be at an angle to the earth's axis equal to the degree of latitude, or 50 degrees, and also at an equal angle to a line passing through the centre of the gyroscope and parallel to the polar axle of the earth.

The performance of the gyroscope and pendulum can only be explained by the assumption that the earth is a revolving sphere. Much of the Zetetic proofs of the world S flatness consist of attacks on the spherical theory. For example, Voliva and his followers argue that an airplane pilot travelling 300 miles an hour would have to 'fall" 60,000 feet an hour in order to maintain his altitude above a spherical world and keep from shooting out in space. Of course, they deny the existence of gravity, and the fact that an airplane maintaining a constant distance from the earth's centre is neither falling nor rising. In addition, the formula on which they calculate the rate of "fall" is not true.

The formula based on the well-known fact that the earth's curvature is eight inches in one mile. In other words if three stakes of equal length are set up on water, at distances of one mile, and a sight is taken over the tops of the two end stakes, two miles apart, it will pass eight inches below the top of the middle stake, one mile away.

But the Volivalites go on to calculate that the "drop" increases as the square of the distance, and therefore that distance squared, multiplied by eight and divided by twelve will give the drop from the horizon in feet for any distance. Actually, there is no formula in spherical trigonometry, which can be used to calculate such a spherical triangle for any desired distance.

That can be proven very easily by anyone who knows plane geometry, for it requires no knowledge of trigonometry to show the error in the formula. If you draw a circle, with its equator and the vertical meridian, giving the north and south poles, and then draw a horizon line at the North Pole, you can prove the fact.
A flyer starting from the North Pole and travelling to the equator would cover 60,000 miles if the globe is 24,000 miles in circumference. The diameter of a 24,000 miles globe is 7,639.69 miles, and the radius is 3,819.845 miles, so the "drop' from the horizon of the North Pole would be the same.

But according to the Zion formula, if you square the distance of the flight, 6,000 miles, you get 36,000,000, multiply that by 8 (the curvature per mile) and you have 288 million, divide by 12 to get the feet, gives 24 million, and divide that again by 5280 to get the miles, and the answer is 4545.4.5 miles, or an error of more than 725 miles. To have the drop which the Zion formula indicates the earth would have to have a diameter of more than 28,588 miles.

The Zion formula also makes no provision for angles in the negative or positive sense, as measured in trigonometry, and if the flyer kept on past the half way mark it would show the drop continuing to increase, instead of decreasing, until, if he returned to his starting point, the formula would show him to be 72,935 miles away in space.

The Voliva prize probably will remain collected unless some future space traveler someday anchors his ship a few thousand miles out in space and takes a movie of a globular world turning on its axis. 'That seems to be the only way the $5,000 can ever be collected.

Universal Gravity is a theory, not a fact, regarding the natural law of attraction. This material should be approached with an open mind, studied carefully, and critically considered. Quantum level objects do not show any signs of gravity. The particle accelerators prove this. They have yet to find any force that works as gravity is described.

A nearly infinite amount of evidence to support a theory has very little weight compared to one single piece of evidence that contradicts it. In point of fact, there might well be a problem with the law of gravity as the universe is not behaving as the current laws of physics would have it behave.

To begin with, the universe is expanding and the rate of expansion is increasing. The law of gravity would expect all the matter in the universe to pull the universe back into single mass, so the universe should either be shrinking or at least the rate of expansion should be slowing down. So either there is new energy being introduced into the universe or we need to re-think our understanding of gravity.

Also the rate that stars on the outer edge of a galaxy orbit the centre of the galaxy is the same as for stars closer to the centre, meaning galaxies maintain their shape. This is contrary to what our normal laws of physics would predict. It is possible that there is a great deal more mass contained in a galaxy than we can currently detect or perhaps our whole understanding of gravity is wrong. For now physicists and astronomers talk of dark energy and dark matter and will likely continue until they relies it is Aether.

ELECTRICITY AND MAGNETISM (flat Earth)
Versus
The Hypothetical "Attraction of Gravitation."
By Lady Blount & Albert Smith. Zetetic Astronomy 1956.

Scientists have long desired to find a physical basis for that which they are pleased to term "the Law of Universal Gravitation." Much better would it have been if they had first sought proof as to whether universal attraction is a fact, or only a mere theory. In many cases the phenomena on which they rest their theory are capable of explanations apart from that theory.

That bodies in some instances are seen to approach each other is a fact; but that their mutual approach is due to an " attraction," or pulling process, on the part of these bodies, is, after all, a mere theory. Hypotheses may be sometimes admissible, but when they are invented to support other hypotheses, they are not only to be doubted but discredited and discarded. The hypothesis of a universal force called Gravitation is based upon, and was indeed invented with a view to support another hypothesis, namely, that the earth and sea together make up a vast globe, whirling away through space, and therefore needing some force or forces to guide it in its mad career, and so control it as to make it conform to what is called its annual orbit round the sun! Theory first of all makes the earth to be a globe; then not a perfect globe, but an oblate spheroid, flattened at the " poles"; then more oblate, until it was in danger of becoming so flattened that it would be like a cheese; and, passing over minor variations of form, we are finally told that the earth is pear-shaped, and that the "ellipsoid has been replaced by an apoid"! What shape it may assume next we cannot t tell; it will depend upon the whim or fancy of some astute and speculating "scientist."

All this of course is said to be due to Gravitation! We have long since given up the theory of gravitation; in fact that theory went with the globular theory which it was invented to support.

We think that the phenomena of celestial motion can be explained by Electricity and Magnetism without having to resort to the theory of universal " attraction " of bodies for each other ; especially attraction at such enormous distances as the astronomers postulate. In short, Zetetics agree with Sir Isaac Newton, that "action at a distance" is impossible without some connecting medium : and that, therefore, bodies at a distance can only act upon each other through the ether, and the electric and magnetic currents which are set up in that subtle substance.

The action of the magnet is, however, supposed to be a proof of the possibility of two bodies "attracting" or pulling each other together from a distance; but when this proof is examined it will not bear this interpretation. If we stand on London Bridge we may sometimes see a boat approach the bridge, by the mere action of the wind or tide. It would be highly unphilosophical to say that the bridge "attracted" the boat; and it is equally unphilosophical to say that-the magnet "attracts" the needle or any other body. As the boat is carried towards the bridge by the action of the tide, or the currents acting directly upon it, so the needle is deflected towards the magnet by the magnetic currents which act upon it.

The magnet, because of its internal arrangement, simply has the power to decide the direction of those currents. When Mr. Adams, or Le Verrier in 1846, discovered the unknown planet Neptune, through the perturbations of the neighbouring planet Uranus, it was, therefore, no proof, as is commonly supposed, of the universality of the Law of Gravitation; for the perturbations of Uranus might be accounted for by electric currents set up between the two planets as they approached each other.

If we were to sit in the telegraph office on this side of the Atlantic, and watch the perturbations of a magnetic needle when a message is being sent across the water, it would not be considered very scientific or philosophical to suppose that some needle on the further side of the ocean was "attracting" or *"pulling"* at the needle on this side I Would it? It would be a much simpler explanation of the phenomenon to say that the magnetic currents set in motion on the one side affected the needle on the other. This is the explanation respecting currents on the earth; and it is the explanation which is given in the case of "wireless telegraphy." But when the philosophers get among the stars with their supposed immense distances, they have to conjure with the word "Gravitation," in spite of all its infinite perplexities, to account for a simple phenomenon. Is this scientific?

A book has lately been published, entitled: *"Aether and Gravitation."* It is a suggestive and well-written book ; but before trying to find out either the cause of gravitation, or its basis as a universal law, it would have been better to have examined whether there really exists such a universal force of attraction, or " pulling together " of particles, as is so commonly assumed. If Mr. Hooper's book proves anything, it really proves that there is no need for any such theory of gravitation; and it may be possible that he has intended to prove this, while at the same time using the old terms or phrases connected with that theory, so as not to excite the opposition of scientists who are still wedded to such an unphilosophical notion as "action at a distance."

But this theory, at the outset, is taken for granted, as is also the globular theory of the earth and its supposed motions. In fact the author in another place seriously sets himself to enquire as to what is the "cause of the earth's diurnal motion"!

Would it not be more logical to first enquire whether the earth really has any such motion? We think so. Astronomers have long been puzzled to account for the earth's supposed diurnal motion. They have no idea what causes it. A primitive impulse will not suffice, as it would require a *continued* and *continual* impulse to equalize the "attraction" theory: and so they have invented what they call a centrifugal as well as a centripetal force. But these "forces" only exist in the brains of astronomers and their disciples.

It would puzzle the wisest of them to give an unanswerable proof either that there are any such "forces," or that the earth has any diurnal or orbital motion arising there from.

Both of these unproved and unprovable theories hinder Mr. Hooper from coming to right and logical conclusions, and so they spoil his book. These theories have beclouded the brightest intellects which have tried to solve the "riddle of the universe." Zetetics want something simpler, something more in harmony with facts, experiments, and general observation; and we are persuaded that the connected and kindred forces of Electricity and Magnetism afford to us all the proof which we need.

The forces of the universe are one: or rather, they are derived from one source, and so are transmutable. They are therefore practically the same, whether applied to things terrestrial or things celestial. To illustrate these we will quote from a current number of the *Tramway and Railway World,* in an article under the above heading. The article of course deals with the practical application of Electricity.

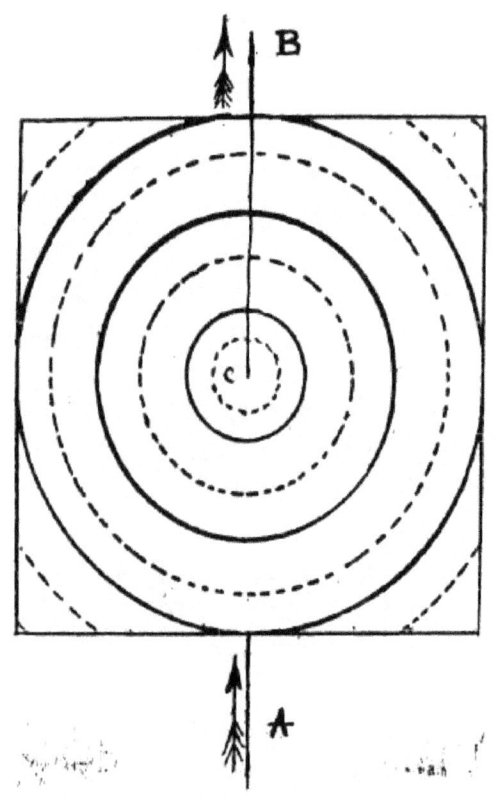

Diagram IV." When a current flows through a straight wire A B, [diagram IV.], a magnetic field is produced around it.

The character of this; field is shown in the figure over the case when the current is flowing upwards through a vertical wire. When the current is flowing Diagram IV downwards, the field is of exactly the same character, except that the lines of force run in the opposite direction round the wire."

Now 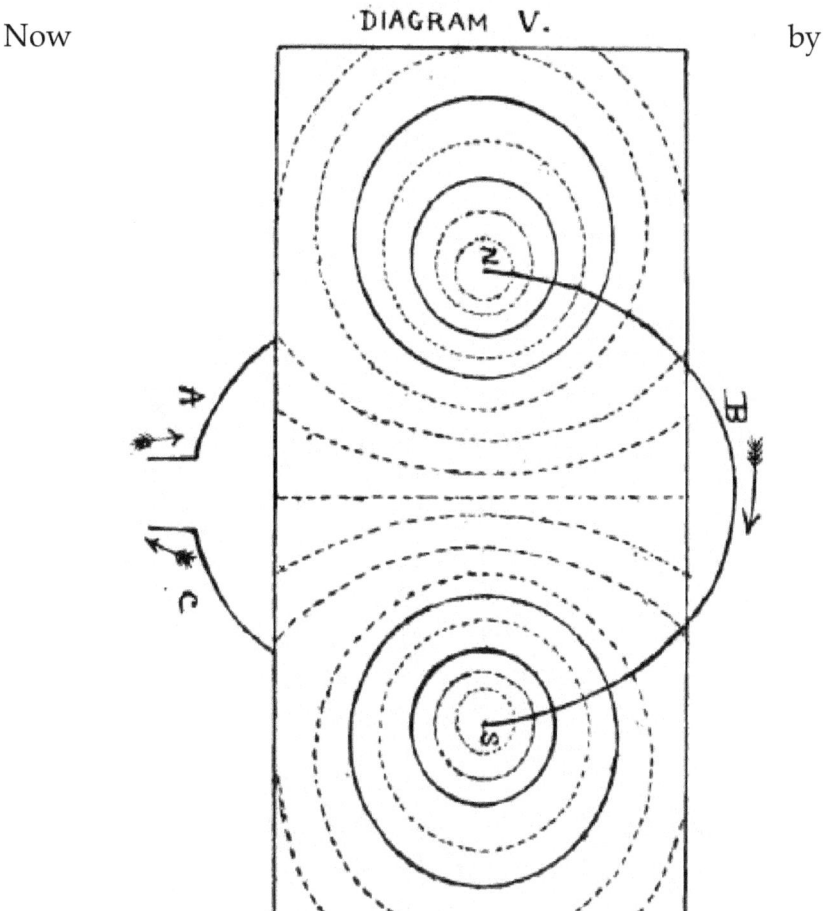 by

reversing the current as is stated above, we get a similar magnetic field with the lines of force going around in the opposite direction. We may take diagram IV. to represent the magnetic field in the northern circle, with (c) as the North Centre.

But instead of reversing the current and so altering the direction of the circular lines of force, we may take a second field to represent the southern circle with the lines of force going round A B C and so from B to C in the opposite direction, as represented in diagram V.

This will illustrate the currents flowing around the South Pole. In the diagram (No. V.) a circular current is represented as going along the wire from A, through N, to B in one direction, and so on from B to C through S in another direction. The electric current thus going in an opposite direction through the wire, at S, from that at point N; the circular lines of forces, or magnetic currents, travel around S in an opposite direction to those which travel around point N; and thus we have an illustration of the two great currents which circle respectively around the North and South magnetic "poles." These great currents meet in a middle and neutral line, or zone, called the Equator, and interlock like the cogs of two connected wheels working together in harmony.

This we will illustrate in Diagram VI.
The rules of the universe are simple. An atoms position in a system is based upon its density in relation to the surrounding densities and the changes in magnetism, electricity and temperature. Density is the most important function in determining the position of an object. Density is the vibration intensity within a volume in relation to the density of the surrounding medium. They suppress Aether Physics because it makes a complete mockery of Mainstream Pseudo Physics. If the Truth were known about how Nazi Aether Craft (UFO'S) worked from the 1940's we would not have this system of science today.

Hollow Earth and Gravity

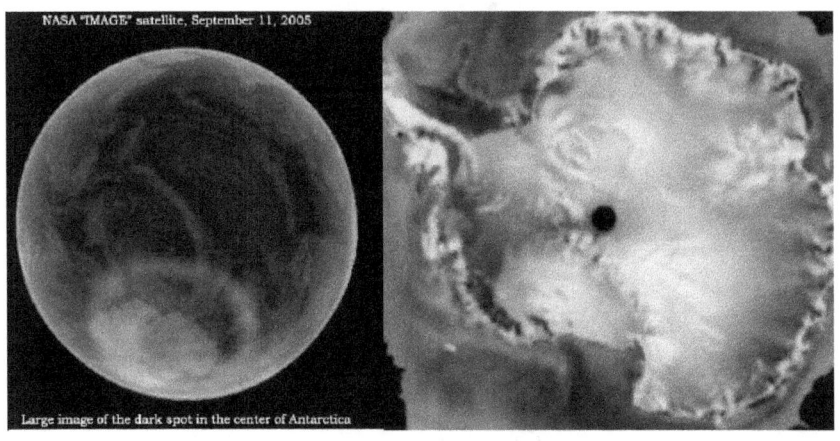

The Aurora Austrais (Southern Lights) emerging from the hole in the South Pole, images can be seen on NASA's own website.

http://www.viewzone.com/hollowearth.html
http://www.nasa.gov/mpg/105410main_FUV_2005-01_v01.mpg
Images from this video

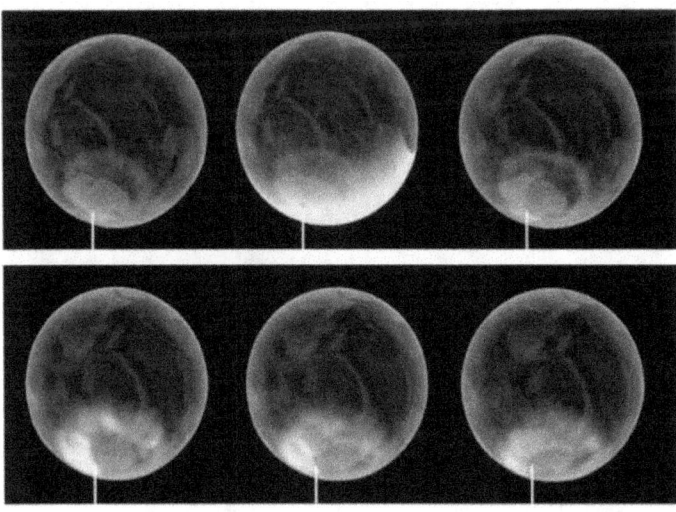

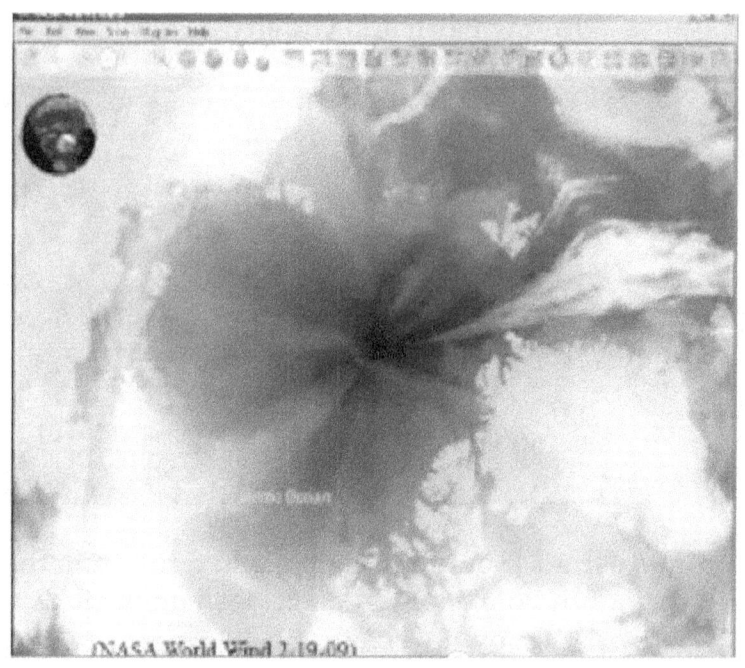

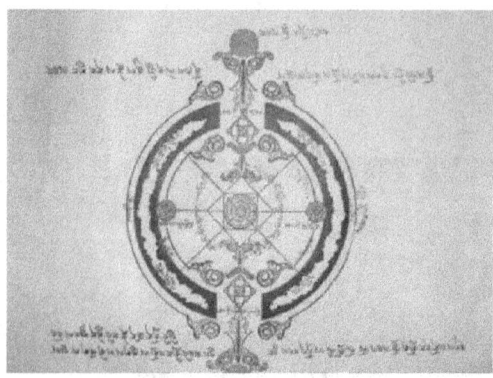
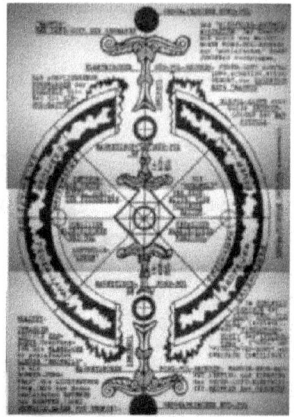

The image above is a depiction of the inner earth from Tibet the Germans took on one of their artifact confiscation expeditions.

Further on you can see an updated version the Germans completed. The German Nazi Regime in the 1930's and 1940's detailing their secret underground bases located on the continent of Antarctica, the most interesting of them being (image below) the one that details in great depth our Earth's hollow interior and the lands that exist there.

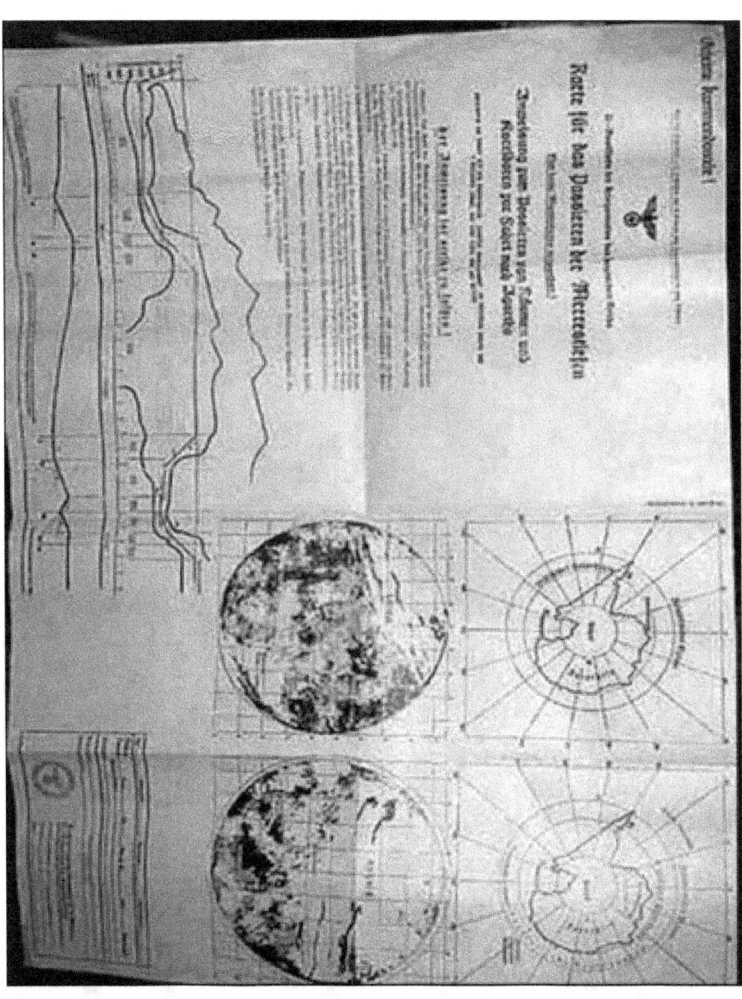

Please do some research into Admiral Byrd's Diary and the forces the USA sent to battle the Germans after the war in Europe finished. See if you can work out where "New Berlin" is located.
https://www.youtube.com/watch?v=BpfxNC4X6xk&lr=1&feature=chclk
The above information can be found here.

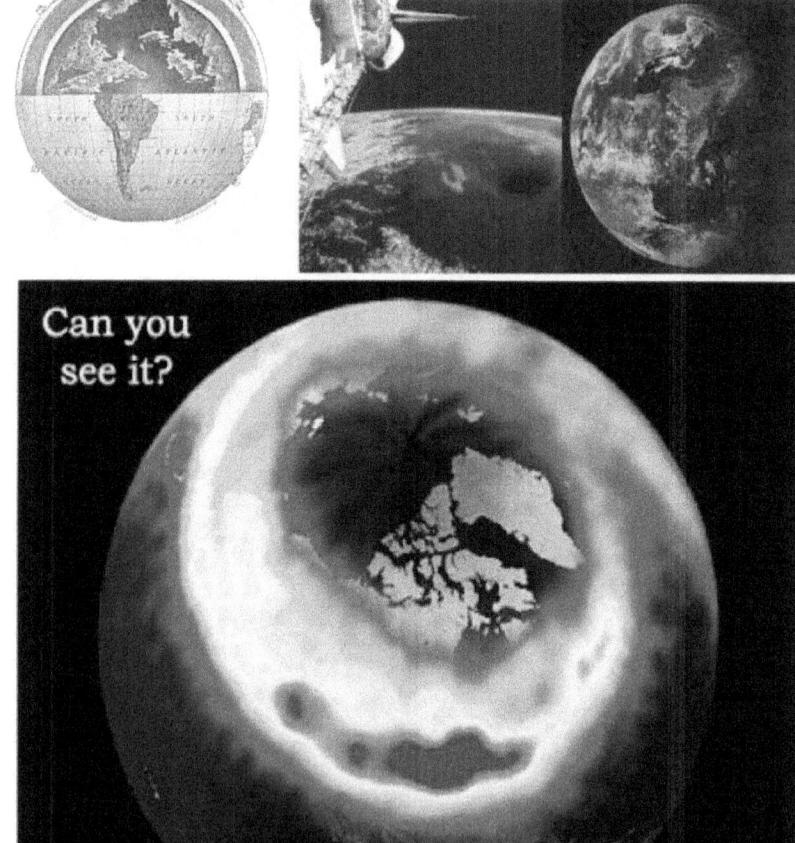

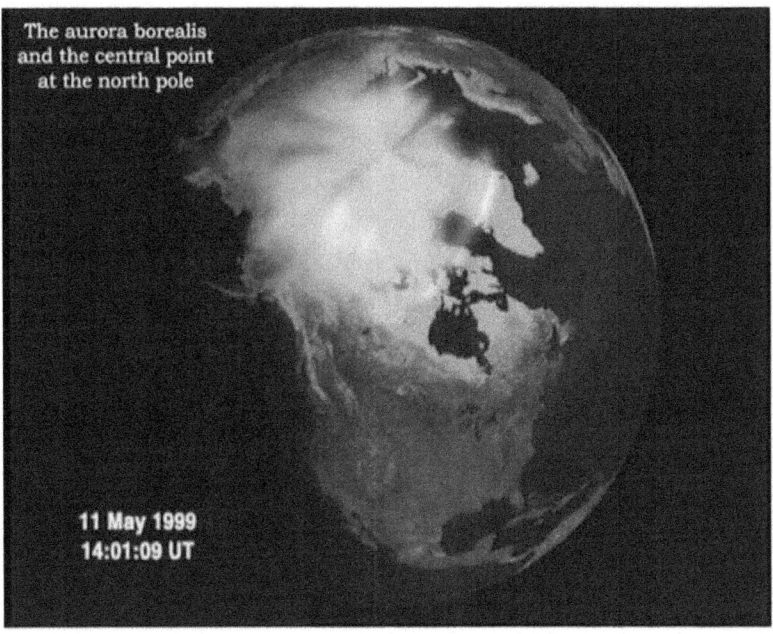

Excerpts from "THE HOLLOW GLOBE OR THE WORLD'S AGITATOR and RECONCILER. A TREATISE ON THE PHYSICAL CONFORMATION of THE EARTH". Presented by M. L. SHERMAN, M. D And "Written by PROF. WM. F. LYON. 1871. (Gravitation Hollow Earth)

Gravitation, to which allusion already has been made, would seem to require more than a passing notice, for as a power or force it has received a large amount of consideration from scientific men of great eminence, and when the discovery was made, that such a power had an existence, and operated in accordance with fixed laws, science seemed to take a long stride, and marched on thence forward with more rapid pace. But, we are of opinion that gravity has been rated too high in the scale of those forces that appear to have been brought into activity, and exerted so prominent an influence in the production of all visible, tangible things that exist: that more, very much more, has been placed upon the shoulders of gravity, than this subordinate, dependent comparatively inactive power is able to bear.

We think it will be ascertained that gravity can hardly be considered to exist as an absolutely active, elemental force, but rather a concomitant of the grosser forms of matter, and that its power is entirely dependent upon the existence, and regulated by the density or peculiar quality of the matter over which it seems to exert an influence. Gravity as a force evidently has no attachment to, or affinity with, or power over material particles that become more etherealized than our atmosphere, and although it seems to act with such potency, upon grosser forms under certain conditions, yet if it may be termed an absolute force, it occupies but a low and very subordinate position.

We have remarked, that upon this elemental structure, the physical globe, all material substances were more or less active in accordance with their grossness, and as we have said, gravity only keeps company with the more gross particles, and it will be seen as activity ceases, gravity usurps its authority and assumes control; for this sluggish force keeps little company with the more active elements. It has little control over the atmospheric particles, scarce any, over vapour cold, in its heated condition, none, and caloric is entirely beyond its reach. But when we travel into the realms of inactivity, and the nearer we approach the state of quiescence, we call death, the nearer we get to the realm where gravity holds his universal reign, because, in this gross department of nature, this force seems to exert its influences, and affinities with the elementary substances.

So when we approximate conditions of matter which is more etherealized, gravity exerts no control; its domain evidently finds an impassable boundary, and its power, a limit beyond which it cannot go.
Again, it would appear evident that if gravitation exerted a controlling influence over this great accretion of material atoms composing the earth, which formed it into a globe, and if it controls the dewdrop, and forms that into a globe also, and the law is so universal as to reach the two extremes, then it ought likewise to form all accretions of fluid or plastic materials, into globes.

But, this is by no means the case; for, in every instance where gravity exerts any control over fluid or plastic materials in any considerable quantities, upon this earth, it universally flattens them or gives them a plane upper surface. Gravity is evidently but a puny, feeble arm of those universal Electro-Magnetic forces that pervade all nature, which is provided for the purpose of reaching out and conducting all ponderous bodies that are gross and inactive, and inclined to rest, to a place upon the bosom of the earth, where they may repose until acted upon by some other superior power. It simply acts in this subordinate capacity, and here may be found the limits of its influence.

the views of Sir Isaac Newton, thus acknowledging, that gravitation was only a proximate cause, and that it was not by any means sufficient of itself to produce the movement of the heavenly bodies, and that it required continually from first to last, the eternal activities of a God behind this force, in order to enable it to produce the planetary movements, and continue the varied evolutions of the universal worlds. Newton requires the re-excitation of motion, or a special operation of Divine power to act upon that force which he had instituted to perform this part of his labour.

This shows that he whom they denominate the Infinite architect, was unable to construct a self-moving machine, or one that would generate the requisite forces for its own propulsion, but, that he has left the machinery in that imperfect condition, in which it requires his constant and unremitting attention to keep it in motion, and that vis inertia, gravitation, and motion, are not sufficient, unaided by Infinite power, to move the worlds.

It will be discovered that both light and warmth being active, positive conditions, are produced by vibrations, or frictionizing activities that continually exist in the fluid essences found upon one-half the earth's surface, caused by coming more immediately under the influence of nature's great Voltaic battery. In order to make this battery effectual in producing the required results, there must be an element existing upon the planet that will assimilate to those where the battery is erected. It is well known, that in telegraphing successfully, the elements at the point of reception must be undisturbed, and in quite a similar condition to those where the battery is erected, which seems to cause the vibration of the electric fluid that transmits the message, as any change or disturbance in the elements entirely prevents the harmonious activities of the instruments, and of course, failure ensues.

We trust then, the reader will learn that the quantity and quality of light, as well as heat depends almost exclusively upon the conditions of the several planets, and the various changes are evidently produced by changes upon them, instead of the sun which acts as the great central Electro-Magnetic battery.

Darkness and cold being inactive and negative, or an absence of those violent frictionizing activities that produce the more positive conditions of light and heat, of course, occupy that half of the globe directly opposite the great magnetic mantle, and these two different conditions are continually changing places from east to west, as the earth rolls upon its axis in the other direction.

The fluid elements within the limits of the atmosphere and contiguous to our globe, upon that side which looks toward the sun, and which are enveloped within the great positive covering of light, being evidently excited to a vibratory frictionizing activity productive of illumination and caloric, as a matter of course, assimilate more nearly to the character of the great positive globe in the centre. While the same elements upon the other side that look out upon the vast, cold, dark, negative space where inactivity reigns, become torpid or inactive, ceasing to a very great extent, their rapid, frictionizing movements. They become cold and dark, assimilating to the negative conditions that exist in the unlimited regions of space, outside the influences of any planetary bodies.

There cannot be a shadow of doubt, but a higher, advanced condition of a world is far superior to a lower, more undeveloped condition; and their projectors and builders must have been capable of carrying out the grand purposes for which worlds are constructed. So, we shall discover, the inhabitants of the exterior planets must be blessed with perpetual day and eternal summer; for there can be no influences we can discover, that would produce the rigors of severe winter, or the scorching heats of a tropical season, where all elements are brought up to that harmonious and elaborated condition of self-dependence that assimilates to the character of the central sun.

Violent extremes of heat and cold are evidently produced by an inharmonious condition of antagonistic or positive and negative elements; and doubtless where those elements are properly evolved and equalized, a happy equilibrium of temperature will prevail.

At the present period in our earth's history, we find great diversity of temperatures in consequence of the unelaborated condition of the elements; at times, and in places, we have too large a supply of the magnetic, and of course correspondingly warm weather; on the other hand, wherever and whenever electricity predominates, it necessarily produces cold in excess.

But, we entertain an unbounded confidence, that those powers who are able to construct an infant world, and manage its own affairs successfully until it attains its majority, gets from under parental influence, becomes independent, and is able to supply its own wants, will be abundantly competent to preserve an equilibrium of the various forces necessary for its continuance, until it accomplishes all its ultimate objects, and to the fullest extent, the original design of the projectors.

Although this portion of the subject is by no means complete, yet we may possibly venture some further reply to the early inquiry that has presented itself to the mind of the reader, when told of a beautiful world in the interior of the spherical shell which we inhabit. A world, too, far more elaborated, and in a more highly finished condition, than this exterior surface which we occupy. The query of course relates to the manner of obtaining illuminating and warming influence, in the absence of the great central magnet of our solar system. We simply remark concerning this matter, that all will become plain, when it is conceded that worlds possess the latent powers within themselves, which enable them to generate those elements, when sufficiently advanced or unfolded.

The interior surface being in a more highly developed condition than the exterior, it has become already capable of generating and producing its own light and warmth, upon the same principle as those planets that are entirely beyond the lighting and warming influences of the sun. The brilliant displays of aural lights that are so frequently beheld emanating from the Arctic Circle, have thus far baffled all attempts of scientific minds to unfold their mysteries; and these phenomena remain to-day, as they ever have, entirely inexplicable. Although they sometimes light up a great portion of the northern hemisphere with unequalled beauty and grandeur, with their softened mellow scintillations, yet all the causes that produce their glories, are shrouded and concealed from the minds of men, in the darkness of Egyptian night.

Very many observations have been made by men of learning, in order to penetrate this mystery, but, as yet, they have resulted in very little that would explain the philosophy of the aurora borealis. They have learned some few facts in connection with those magnificent displays, and there the matter rests, as far as science is concerned. We conclude it is not too much to say, and we venture the opinion, based upon analogical reasoning, that the interior surface of our globe is already unfolded to a condition quite as high as the exterior of the outer independent planets. That the beautiful aural and magnetic lights, and genial warmth, are all produced by the more advanced inherent powers existing within this shell, and that the aural polar lights are to a great extent generated by powers and elements that exist in and emanate from the interior world.

We hesitate not to assert, if there was no such beautifully unfolded inner world connected with the Polar Regions, there would be no such grand illuminations in the north, or in the south, to awake the sublimest emotions in the mind of every beholder.

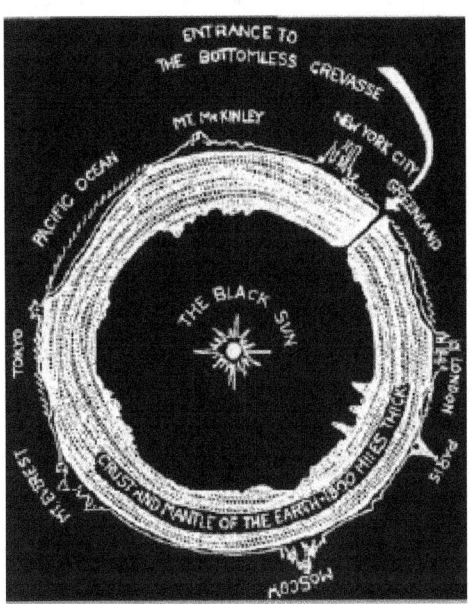

I believe what is meant "by the more advanced inherent powers existing within this shell, and that the aural polar lights are to a great extent generated by powers and elements that exist in and emanate from the interior world" is the BLACK SUN.

I had been reading an article saying that gravitational force cannot be electromagnetic as an electromagnetic does not attract all elements, for example non-ferrous metals. Well you can make an electromagnet attract non-ferrous metals as I have a book on how to do this and have made such an electromagnet. This book is called "Design, Construction & Operating Principles of Electromagnets for Attracting Copper, Aluminum & other Non-ferrous Metals" by Leonard R. Crow 1951.

The following information is an excerpt from **"The Awesome Life Force" by Joseph Cater 1984**. I enjoyed reading his book very much especially the chapters on "Incontrovertible Flaws in the Theory of Relativity" and "Incredible Flaws and Discrepancies of Orthodox Science." It is well worth a read. NASA as well as other interests has gone to great lengths to cover up the real findings of the space program. Such findings provide undeniable proof the most celebrated theories and concepts of conventional physics are completely erroneous. Despite all efforts to camouflage their program, leaks and slip-ups did occur. These leaks opened the door to many incredible revelations for any intelligent researcher with the necessary dedication and perseverance to take full advantage of the situation. Such findings will now be summarized. The following Items are not mere speculation but are realities with a great wealth of factual evidence in conjunction with logical analysis to substantiate them.

1. Methods of propulsion of other than rockets were employed during critical stages of the Apollo missions. The space ships could not have carried sufficient fuel for the astronauts to reach the moon and return because of the moon's high gravity. The lift-off from the earth with rockets was part of the great NASA cover-up.

2. The earth (as all planets are) is hollow with a great egress, or entrance, into the earth's interior that is hundreds of miles across. It is located in the north Polar Regions just south of the North Pole. Earlier satellite pictures of the earth show this entrance quite clearly. Subsequent pictures released by NASA were doctored to obliterate any evidence of such an entrance. Apparently, they neglected to do this with the earlier releases.

3. For years prior to the Apollo missions, NASA had space ships capable of interplanetary travel at its disposal. These ships employ fuel less propulsion systems similar to that of the highly publicized UFO (The principles will be analysed later on in this treatise).

4. Gravity effects are produced by a highly penetrating radiation in the electromagnetic spectrum. It can be produced by mechanical means and used as levitating beams, as well as for a very effective method of propulsion. NASA has had such devices for many years.

SUMMARY OF PART II

Many of the facts and principles introduced in this part have not appeared in print before. It was shown that science has failed to explain any of the common everyday phenomena, such as tides, which are taken for granted. Further evidence was supplied showing that our planet is indeed hollow, with far better living conditions in the interior than exist on the outside. Additional steps were taken toward resolving the mystery of gravity. It was shown that popular theories in the field of cosmology are as devoid of sound thinking as many other debunked academic theories. Also, a new insight into the nature of the ethers was introduced for the first time, laying the groundwork for a deeper understanding of a wide range of phenomena to be given in Part III and IV.

Greater surprises are in store for the reader in Part III. Some of the ideas introduced in Parts I and II will be developed still further and new concepts given, embracing the entire field of physics. This will also include the explanation for geomagnetism not discussed in Part II, since more groundwork in the nature of magnetism needs to be presented for a proper treatment of this subject.

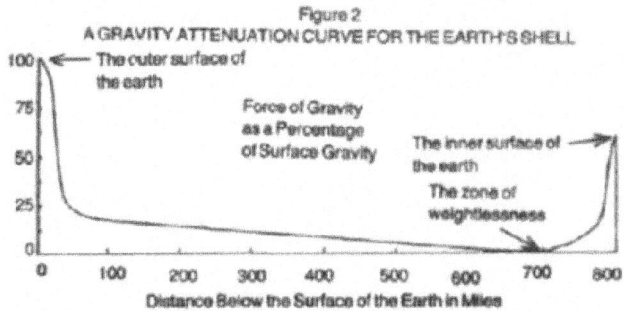

Figure 2
A GRAVITY ATTENUATION CURVE FOR THE EARTH'S SHELL

In fact, the U.S. Government sponsored experiments as early as 1958 that proved such devices feasible. Gravity drops off rapidly in the first 25 miles below the earth's surface due to the limited penetrating ability of gravity-inducing radiations. From that point downward the rate of decrease becomes progressively less until it drops to Zero 700 miles below the outer surface.

The force of gravity begins rising again proceeding towards the inner surface which is at the 800 mile depth. At the inner surface, the force of gravity reaches a value which is somewhat less than that on the outer surface of the earth.

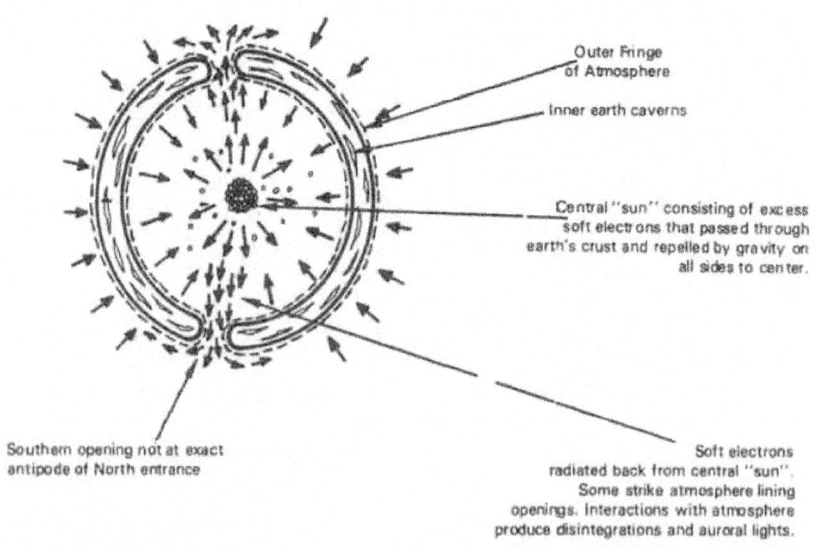

Figure 3
CROSS SECTION OF EARTH TAKEN THROUGH NORTH AND SOUTH OPENINGS

Large openings prevent excessive accumulation of soft electrons inside the earth. They function as an exhaust system for the excess particles to escape into outer space. Were it not for this there would be a steady build-up of heat throughout the inner earth and the crust with disastrous results. Radar pictures taken of Venus show enormous openings at almost the exact antipode of each other and they are round. Being closer to the sun, Venus would require much larger openings. The surface is protected from the extreme radiation from the sun by a very extensive mantle of water vapour. This belies the claims of conventional science which state Venus has a surface temperature of about 1000 degrees Fahrenheit with sulphuric acid clouds. This is consistent with other false pictures they paint of the universe.

The Globe from a Different Perspective

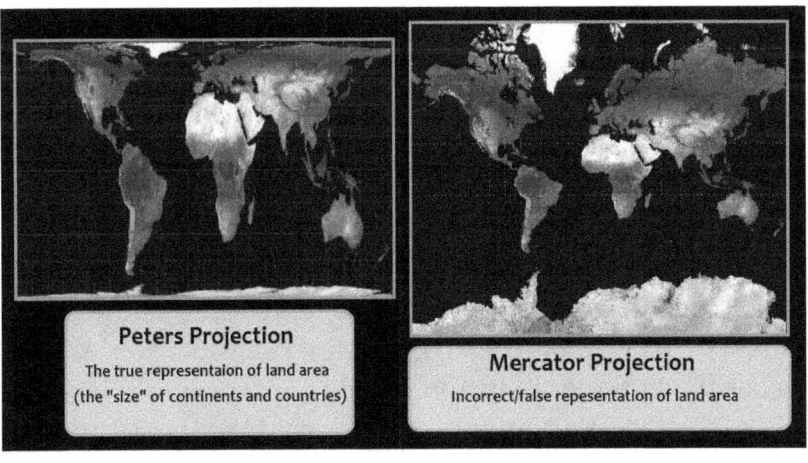

Peters Projection
The true representaion of land area
(the "size" of continents and countries)

Mercator Projection
Incorrect/false repesentation of land area

As you can see the Peters Projection map is a true representation of land area and size of the continents whereas the Mercator projection that we currently use and is accepted is clearly way out of proportion. This should clearly tell you that even though what we use now is the excepted version it certainly does not mean it is in anyway correct. This goes for anything that is currently accepted as fact. Another point of interest is the southern polar region and how it covers the entire bottom of the map as if went around the world and the northern polar region only covers a small section.

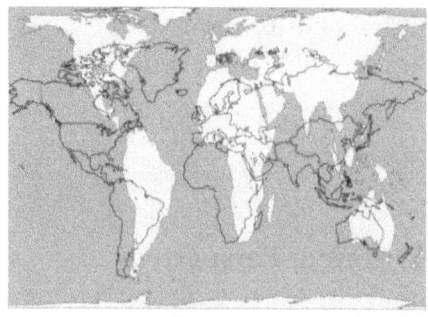

I wonder why that is?

Look at the dramatic difference in the size and global positions of the landmasses. How could they allow such a distorted view of the world become the accepted norm.

Therefore, we have seen the Flat Earth Theory and Hollow Earth Theory and arguments associated with both, so which one is correct. One thing is sure there is not a solid globe with molten core as mainstream pseudo-science teaches us. You would be foolish to think the world is a solid globe this is a certainty. Think what the third option is.

Ring of the Gods, the Stargate

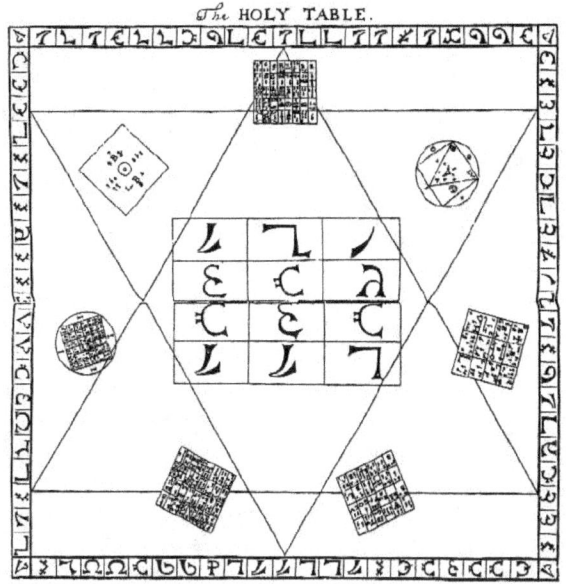

The Star Gate "Scrying Table" of Dee and Kelly served as a portal to other universes. The Language of light (software) Continued. Modern experiments reproduce these results.
http://www.bibliotecapleyades.net

Above: Stargate Circa 1923? Take note of the symbols on the stargate as they are the planetary symbols connected to Magic Squares or as they are also known as the Planetary Squares.

Every Planet and every Element of the periodic table has its own magic square and symbol associated to it. The science of Alchemy is the transmutation of one element to another and the information has been suppressed so we can all be slaves in a debt based system based on corruption. There is no need for money, mining or oil exploration when you can just turn one element into another.

Stellar ship from the tomb of Ramses VI (top), wormhole (bottom).

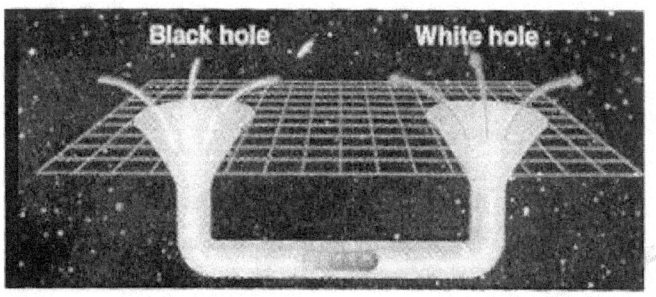

Is the shape of the Egyptian ship of the gods a primordial design for a wormhole?

A simple explanation on how the stargate works would be comparing it to the old rotary dial telephone but having symbols instead of numbers. As you can see above the symbols overlay on the magic square and this gives you the dialing sequence for that planet. Egyptian symbolism has been intentionally misinterpreted by people I have an utmost of contempt for.

The following images of magic squares and symbols are from. THE MAGUS, OR CELESTIAL INTELLIGENCER; A COMPLETE SYSTEM OF OCCULT PHILOSOPHY BY FRANCIS BARRETT, F.R.C. Professor of Chemistry, natural and occult Philosophy. LONDON: PRINTED FOR LACKINGTON, ALLEY, AND CO.

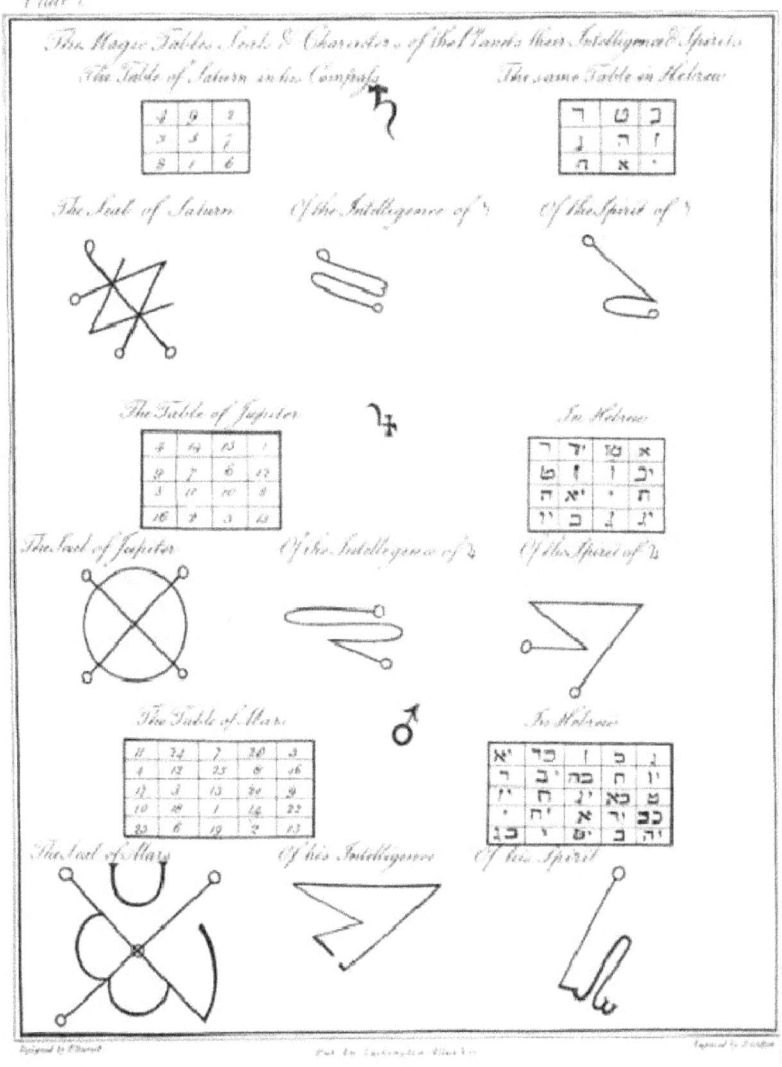

TEMPLE OF THE MUSES, FINSBURY SQUARE. 1801.

The magic seals and talismans are the symbols for the stargate and transmutation of elements.

The following images are from that great book "A Primer of Higher Space (The Fourth Dimension) By Claude Bragdon 1913. This is a simple explanation of magic squares and cubes to help you understand more clearly. The pseudo physics we have today needs to be replaced.

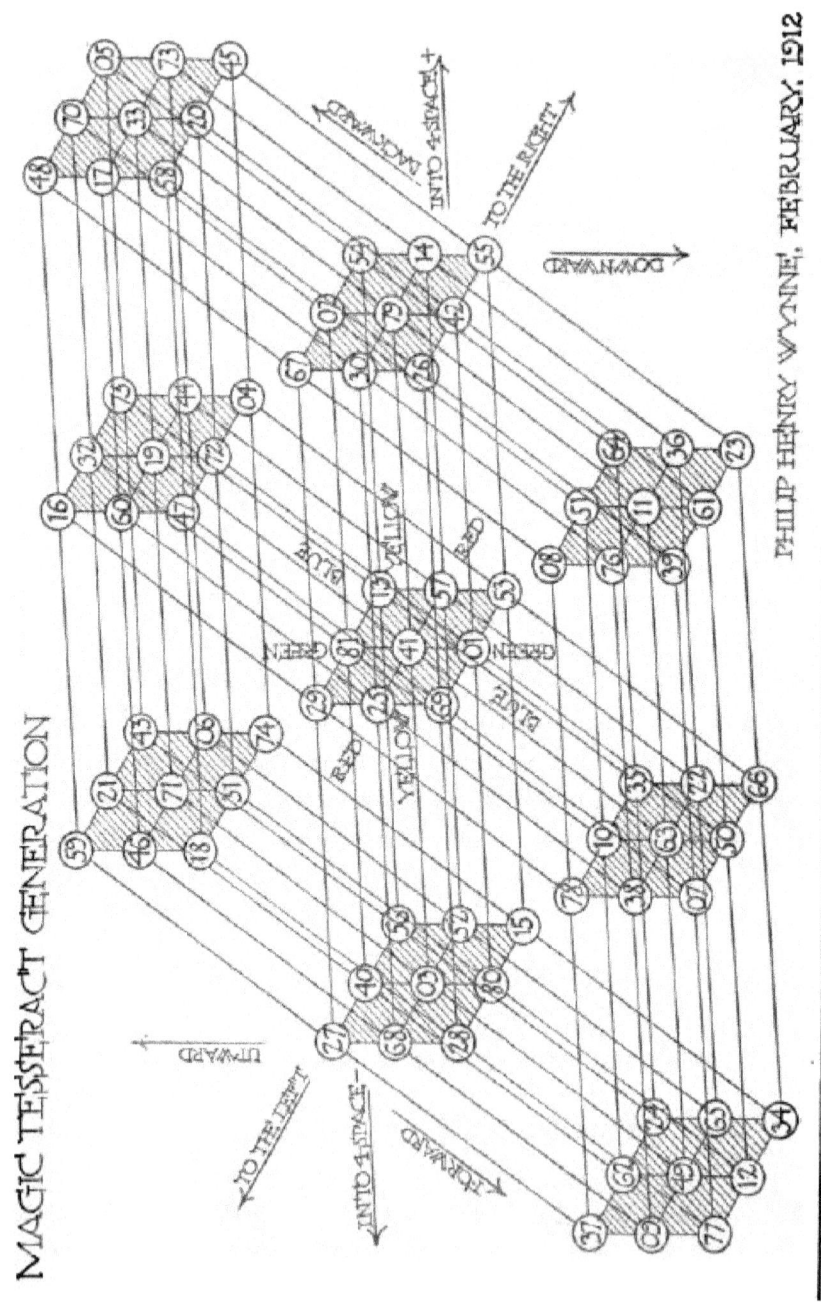

INTRODUCTION TO THE MAGIC TESSERACT.

THE UNTRAINED MIND FINDS SOMETHING ESPECIALLY SATISFACTORY AND SECURE IN NUMERICAL RELATIONS. IT EASILY BELIEVES THAT WHAT IS TRUE OF 4, 17, OR 1001 UNITS MUST BE TRUE OF 4, 17, OR 1001 APPLES—MILES, OR MEN. THE AUTHOR ACCORDINGLY INVOKED THE MATHEMATICAL GENIUS OF PHILIP HENRY WYNNE TO FURNISH HIM WITH SOME INEXPUGNABLE ILLUSTRATION, FOUNDED UPON THE PROPERTIES OF NUMBER, OF AN OPEN ROAD FOR HUMAN THOUGHT INTO THE FOURTH DIMENSION.

THE RESULT OF MR. WYNNE'S DIVING INTO THE DEEP WATERS OF MATHEMATICS WAS THE PRODUCTION OF THE MAGIC TESSERACT, A PEARL, WHICH ANYBODY IS FREE TO EXAMINE AND ADMIRE, BUT ONE WHICH ONLY A MATHEMATICIAN CAN PROPERLY APPRAISE. IF THE READER WILL FOLLOW THE ENSUING EXPLANATION STEP BY STEP, VERIFYING THE RELATIONS NOTED, HE WILL BE ABLE TO PARTICIPATE, WITHOUT DEEP KNOWLEDGE OR HARD LABOR, IN THE ASSURANCE OF THE UNDERLYING REALITY OF HYPERSPACE WHICH COMES TO THE MATHEMATICIAN OF AN OPEN MIND, AS A RESULT OF HIS RESEARCHES.

A MAGIC SQUARE OF FOUR

1	14	15	4
8	11	10	5
12	7	6	9
13	2	3	16

FIG 1.

THE ACCOMPANYING FIGURE REPRESENTS A MAGIC SQUARE OF 4 MADE BY A WELL KNOWN METHOD. THE READER SHOULD VERIFY THE FOLLOWING RELATIONS, FOR THEY ARE NOT TRIVIAL IN CONNECTION WITH WHAT COMES LATER.

EACH HORIZONTAL AND EACH VERTICAL COLUMN ADDS 34. EACH DIAGONAL ADDS 34. FOUR CORNER CELLS ADD 34. FOUR CENTRAL CELLS ADD 34. TWO MIDDLE CELLS OF TOP ROW ADD 34 WITH TWO OF BOTTOM ROW; SIMILARLY WITH MIDDLE CELLS OF RIGHT AND LEFT COLUMNS. GO ROUND THE SQUARE CLOCKWISE: 1ST CELL BEYOND 1ST CORNER + 1ST BEYOND 2D + ... 3RD + ... 4TH = 34. EACH CORNER SET OFF BY HEAVY LINES ADDS 34.

TAKE ANY NUMBER AT RANDOM; FIND THE THREE OTHER NUMBERS CORRESPONDING TO IT IN ANY MANNER WHICH RESPECTS SYMMETRICALLY TWO DIMENSIONS AND THE SUM OF THE FOUR NUMBERS WILL BE 34.

A MAGIC CUBE OF FOUR

THIS DRAWING REPRESENTS A MAGIC CUBE OF FOUR CONSTRUCTED BY A PURE EXTENSION OF THE SQUARE METHOD OF FIG 1. PLATE 22.

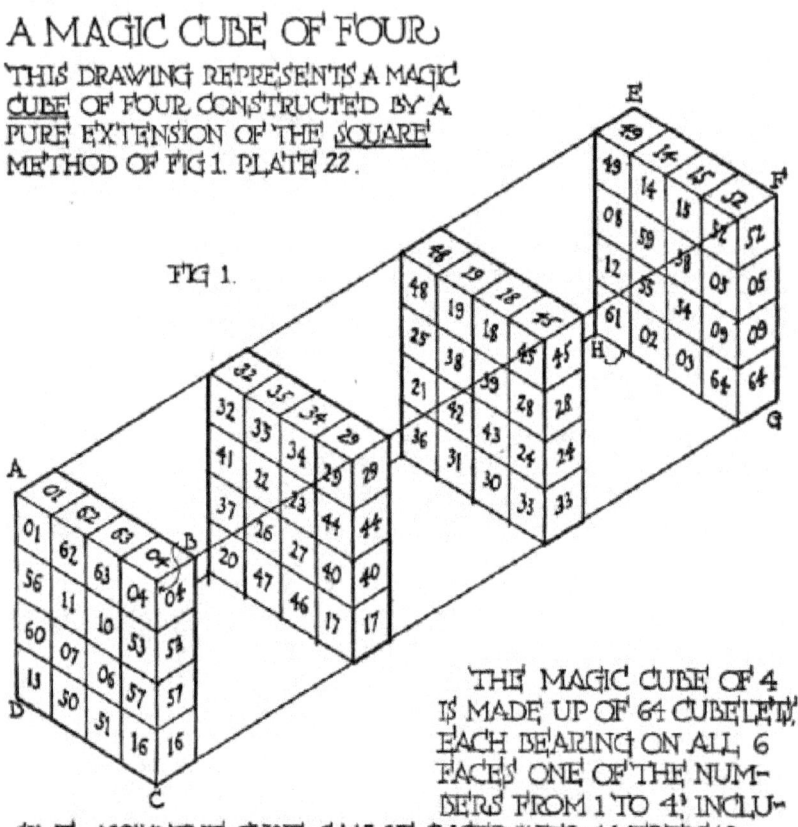

FIG 1.

THE MAGIC CUBE OF 4 IS MADE UP OF 64 CUBELETS, EACH BEARING ON ALL 6 FACES ONE OF THE NUMBERS FROM 1 TO 4, INCLUSIVE. NOW THE CUBE CAN BE SLICED INTO 4 VERTICAL SECTIONS FROM LEFT TO RIGHT AS IN FIG 1, WHICH SHOWS THE SECTIONS SEPARATED SO THAT THE INTERIOR SUMMATIONS CAN BE SEEN.

OR IT CAN BE SEPARATED INTO OTHER 4 VERTICAL SECTIONS BY CUTTING PLANES PERPENDICULAR TO THE EDGE A B—PROCEEDING FROM FRONT TO BACK.

OR THE 4 SECTIONS MAY BE HORIZONTAL, MADE BY PLANES PERPENDICULAR TO A D.

NOW EACH ONE OF THESE 12 SECTIONS PRESENTS A MAGIC SQUARE IN WHICH EACH ROW AND EACH COLUMN ADDS 130. THE DIAGONALS OF THESE SQUARES DO NOT ADD 130, BUT THE 4 DIAGONALS OF THE CUBE DO ADD 130

THE READER SHOULD VERIFY A FEW OF THESE SUMMATIONS IN EACH OF THE THREE SETS OF SECTIONS.

A FURTHER CONSIDERATION OF MAGIC FIGURES

A DETAILED STUDY OF THE MAGIC CUBE OF 4 WILL SERVE TO CONVINCE THE READER OF THE ESSENTIALLY SOLID NATURE OF THE SPACE ARRANGEMENT AND THE PERFECT CONTINUITY OF THE "MAGICAL" PROPERTIES OF NUMBER FROM TWO DIMENSIONS TO THREE.

NOW THESE SAME NUMERICAL PROPERTIES CONTINUE UNINTERRUPTEDLY INTO THE FOURTH DIMENSION. TO FIX THE IDEA MORE CONCRETELY, CONSIDER THE FOLLOWING STATEMENTS BASED ON THE POWERS OF THE NUMBER 7.

FIG 1.
7 2 3 4 5 6 1

FIG 1. REPRESENTS A MAGIC LINE OF 7. OBSERVE THAT THE SUMS OF PAIRS OF NUMBERS EQUIDISTANT FROM 4 IS

$$\frac{7^1+1}{2} \times 2$$

FIG 2.

30	39	48	1	10	19	28
38	47	7	9	18	27	29
46	6	8	17	26	35	37
5	14	16	25	34	36	45
13	15	24	33	42	44	4
21	23	32	41	43	3	12
22	31	40	49	2	11	20

FIG 2 REPRESENTS A MAGIC SQUARE OF 7. OBSERVE THAT HERE THE VARIOUS SUMMATIONS OF 7 NUMBERS GIVE

$$\frac{7^2+1}{2} \times 7$$

IN THE MAGIC CUBE OF 7 THE SUMMATIONS ARE

$$\frac{7^3+1}{2} \times 7$$

IN THE MAGIC TESSERACT THE SUMMATIONS ARE

$$\frac{7^4+1}{2} \times 7$$

THE SUM IN THE CASE OF THE MAGIC LINE IS NOT ANOMALOUS IN THAT $\frac{z+1}{2}$ IS MULTIPLIED BY 2 INSTEAD OF BY 7; IT IS DUE TO THE FACT THAT WE ARE ADDING BUT 2 NUMBERS TOGETHER INSTEAD OF 7 NUMBERS AS IN THE OTHER CASES. IF WE TOOK ALL 7 NUMBERS OF THE MAGIC LINE WE SHOULD HAVE ONLY ONE SUM AND NO "MAGICAL" CORRESPONDENCES.

NOT ONLY SUCH SERIES AS 1, 2, 3, 4 BUT ARITHMETICAL PROGRESSIONS IN GENERAL, GEOMETRICAL PROGRESSIONS, AND OTHER SERIAL FUNCTIONS SUBMIT TO MAGICAL ARRANGEMENTS IN N DIMENSIONS.

A MAGIC SQUARE OF THREE—A CUBE OF THREE

OBSERVE THAT A MAGIC LINE CANNOT BE FORMED OF LESS THAN 4 NUMBERS, OR A MAGIC SQUARE OF LESS THAN 9 [FIG 1].

FIG 1.

8	1	6
3	5	7
4	9	2

A MAGIC CUBE REQUIRES AT LEAST 27 NUMBERS [FIG 2], AND EVEN WITH THIS NUMBER THERE ARE MANY LIMITATIONS DUE TO LACK OF SCOPE FOR NUMEROUS COMBINATIONS POSSIBLE WITH CUBES OF 5, OR BETTER, OF 7.

MAGIC CUBE OF THREE

EACH ROW AND EACH COLUMN OF EACH OF THE 9 MAGIC SQUARES (3 SQUARES TO EACH DIMENSION) ADDS 42.

EACH OF THE 4 CUBE DIAGONALS ADDS 42.

EACH OF THE 6 DIAGONALS OF THE 3 MAGIC SQUARES CONTAINING THE CENTRAL NUMBER 14, ADDS 42

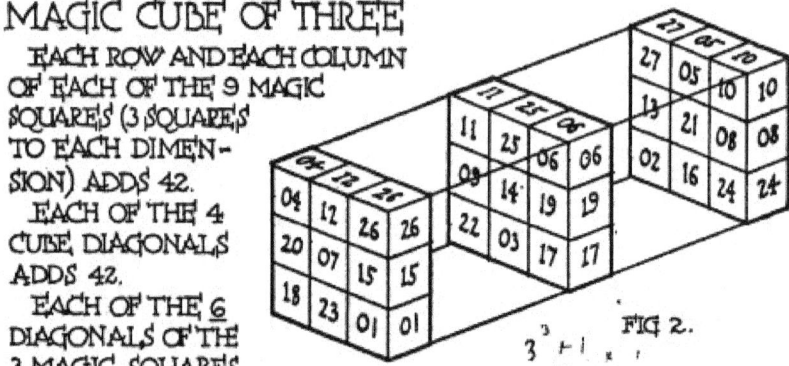

FIG 2.

$3^3 + 1$

ANY RANDOM LINE DRAWN THROUGH THE CENTER OF THE CUBE WILL CUT TWO SURFACE CELLS WHOSE NUMBERS WILL ADD TWICE THE CENTRAL NUMBER 14. ALL THESE LINES ARE "MAGIC" LINES.*

OUR WELL KNOWN NUMBERS 1, 2, 3, 4 ETC. CONTAIN AMONG THEM FOUR-DIMENSIONAL MAGIC PATHS AS REAL AND DEMONSTRABLE AS THOSE IN 2-SPACE.

A MAN WILL OFTEN GIVE "TWO AND TWO MAKES FOUR" AS AN EXAMPLE OF PERFECT CERTITUDE. THESE MAGIC NUMBER PROPERTIES RETORT ON HIM BY CHALLENGING HIS DENIAL OF THE FOURTH DIMENSION WHICH HIS OWN WELL-TRUSTED NUMBERS AFFIRM MOST INSISTENTLY

---

* THIS CUBE OF 3 (AN ODD NUMBER) HAS A CENTRAL CUBELET, WHICH GIVES RISE TO SEVERAL INTERESTING PROPERTIES LACKING IN THE 4-CUBE.

EXPLANATION OF THE MAGIC TESSERACT

THE ATTEMPT WILL NOW BE MADE TO RENDER CLEARLY INTELLIGIBLE MR. WYNNE'S MAGIC TESSERACT. THE READER IS URGED TO VERIFY FOR HIMSELF THE SUMMATIONS PARTICULARLY SPECIFIED AND THE VARIOUS OTHER RELATIONS POINTED OUT. A DETAILED STUDY OF THE FIGURE WILL RESULT IN THE UNFOLDING OF SOME OF THE AMAZING AND BEAUTIFUL INTERRELATIONS THAT LIE OUTSIDE 3-SPACE HORIZONS.

FIRST WE MUST CONSIDER THE SYMBOLS AND CONVENTIONS USED. INSTEAD OF THE USUAL MAGIC SQUARE DIAGRAM OF FIG. 1 THE ARRANGEMENT OF FIG. 2 WILL BE EMPLOYED.

IN FIG. 2 THE NUMBERS MAY BE IMAGINED TO BE ENCLOSED IN CRYSTAL SPHERES SUPPORTED ON A WIRE FRAMEWORK WHICH ALLOWS US TO SEE INTO THE INTERSPACES. THE WIRES MAY BE BENT ASIDE IF NECESSARY, AND ARE SUPPOSED TO BE CAPABLE OF EXTENSION OR SHORTENING AT OUR PLEASURE, THOUGH NOT SPONTANEOUSLY ELASTIC. PLATE 20 REPRESENTS 6 RECTANGULAR PRISMS STRETCHED OUT FROM ORIGINALLY CUBICAL FORM BY THE EXTENSION OF THE BLUE AND THE YELLOW WIRES. THIS DISTORTION IS OF COURSE TO EXHIBIT THE NUMBERS MORE CLEARLY, AND THE PRISMS SHOULD BE CONCEIVED AS PUSHED BACK AFTER INSPECTION INTO THEIR PROPER CUBICAL FORMS.

REMEMBER, THEN, THAT RED, GREEN, BLUE, YELLOW WIRES ARE NORMALLY OF EXACTLY EQUAL LENGTH.

FURTHER (SEE PLATE 20):—

  (R)ED STANDS FOR MOTION TO THE (R)IGHT;
  (G)REEN " " " TOWARDS THE (G)ROUND (DOWNWARDS);
  (B)LUE " " " (B)ACKWARDS (AWAY FROM US);
  (Y)ELLOW " " " IN THE POSITIVE SENSE OF THE <u>FOURTH DIMENSION</u>; INTO (Y)AMAPURA.

THE MAGIC TESSERACT—CONTINUED

NOW TURN TO PLATE 20—CONSIDER THE SPHERE CONTAINING 37; LET IT MOVE TO THE (R)IGHT ALONG THE (R)ED LINE, LEAVING AT EQUAL INTERVALS LIKE SPHERES FOR 62 AND 24. THERE ARE THEN 3 SPHERES, THIS BEING A 3-TESSERACT. OBSERVE THAT 37+62+24=123, THE MAGIC SUM: = [(1+81)÷2]×3, OR MORE GENERALLY, $\Sigma = \frac{n+n^4}{2}$, THE CORRECT SUM FOR 4-SPACE. [FIG 1].

NEXT THE LINE 37, 62, 24, SHALL MOVE DOWNWARD—TOWARDS THE (G)ROUND—ALONG (G)REEN LINES, LEAVING 2 SIMILAR LINES, TO WIT, 9, 49, 65, 77, 12, 34. EACH LINE YIELDS THE MAGIC SUM 123, SO ALSO DO THE 3 VERTICAL LINES. WE THEREFORE HAVE A MAGIC 3-SQUARE [FIG 2].

THIS SQUARE SHALL NOW MOVE (B)ACK-WARDS ALONG (B)LUE PATHS, LEAVING TWO SIMILAR SQUARES AND GENERATING THE FIRST (STRETCHED-OUT) MAGIC 3-CUBE WHICH WE WILL CALL 37-74. NOTE THAT ALL THE BLUE LINES AS WELL AS BOTH THE OLD AND THE NEWLY-GENERATED RED AND GREEN LINES GIVE THE MAGIC SUM 123. [FIG 3].

THE DIAGONALS OF THE MAGIC SQUARES AND OF THE MAGIC CUBES DO NOT ADD 123 EXCEPT WHEN THEY CROSS THE CENTER OF THE TESSERACT. ALL 8 OF THE TESSERACT DIAGONALS DO ADD 123. BOTH THESE FACTS ARE AS THEY SHOULD BE THOUGH THE HIGHER RELATIONS WHICH ARE THE CAUSES NEED NOT BE EXPOUNDED HERE.

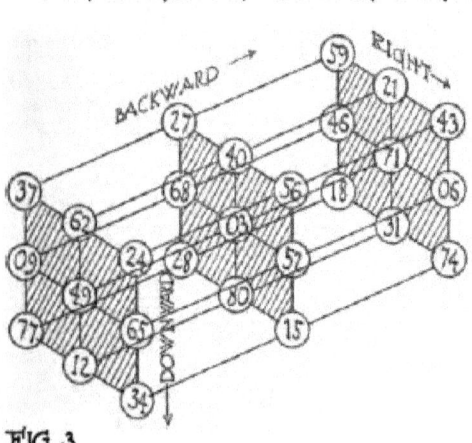

FIG 3.

THE MAGIC TESSERACT — CONTINUED —

NOW WE WILL TAKE OUR LONG LEAP OUT INTO THE DARK.

LET OUR FIRST CUBE, 37-74, MOVE OFF PERPENDICULARLY IN THE + SENSE OF THE FOURTH DIMENSION, LEAVING IN 4-SPACE TWO SIMILAR 3-SPACE CUBES, AND GENERATING THE MAGIC TESSERACT. THE PATHS OF THE CRYSTAL SPHERES ARE (SYMBOLICALLY) REPRESENTED BY THE SYSTEM OF LINES DESIGNATED AS YELLOW.

IF WE REALLY HAVE ACHIEVED THE MAGIC TESSERACT, THEN EVERY RED (MARKED) WIRE (1), GREEN WIRE (2), BLUE WIRE (3), YELLOW WIRE (4), MUST BEAR 3 NUMBERS YIELDING THE MAGIC SUM 123. THIS WILL BE FOUND TO BE THE CASE.

OBSERVE THAT ALL 4 DIMENSIONS ARE PERFECTLY EQUIPOTENT. FOR CLEARNESS THE WIRES MARKED BLUE AND YELLOW HAVE BEEN STRETCHED, BUT THIS MIGHT HAVE BEEN DONE TO THE RED AND TO THE GREEN INSTEAD. WHEN THE TESSERACT HAS BEEN PUSHED BACK INTO UNDISTORTED FORM THE RED, GREEN, BLUE, YELLOW WIRES ARE ALL OF EQUAL LENGTH, AND EACH OF THE 108 WIRES CARRIES 3 NUMBERS WHOSE SUM IS 123.

THIS BEING SO, WHO SHALL SAY WHICH DIMENSION IS MORE _REAL_ THAN THE OTHERS, AND WHY?

THUS IS THE MAGIC TESSERACT GENERATED. LET US NOW SEE IF WE CAN CONFIRM OUR SOMEWHAT ANALYTIC CONSIDERATION BY COLLECTING OUR RESULTS IN A MORE _IMAGINABLE_ FORM. IN SHORT, LET US TRY TO PROJECT BACK FROM 4-SPACE THE CONCRETE ASSEMBLAGE OF NUMBERS WHICH CONSTITUTE THE MAGIC TESSERACT.

WE FIND THAT WE CAN. THE RESULT IS SHOWN IN PLATE 21.

THE TINTING WILL SUGGEST TO THE EYE AN OUTER AND AN INNER CUBE, WHICH ARE RESPECTIVELY 37-74 AND 08-45 OF PLATE 20 (A AND C, FIG 1). BUT THE READER MUST ALSO IMAGINE ANOTHER CUBE, 78-04 OF PLATE 21 (B, FIG 1), HALF WAY BETWEEN THE OTHER TWO (SEE 78, 35, 66, 07, 47, 16, 75, 04, IN THE PROJECTION ALSO).

FIG 1.

THE MAGIC TESSERACT—CONCLUDED

THE READER SHOULD SATISFY HIMSELF BY TRIAL THAT EVERY LINE IN PLATE 20 HAS ITS CORRESPONDING LINE IN PLATE 21. ALSO THAT EACH LINE OF THE 108 IN THE PROJECTION CARRIES THE SAME 3 NUMBERS AS THE CORRESPONDING LINE OF THE 108 IN THE GENERATION.

THE FOLLOWING POINTS ARE IMPORTANT. IT IS PURELY ARBITRARY WHICH CUBE IS PUT APPARENTLY "WITHIN" OR "WITHOUT." THEY ARE REALLY SIDE BY SIDE IN 4-SPACE AND MERELY APPEAR TO BE CONTAINED ONE WITHIN ANOTHER; EXACTLY AS IN A PERSPECTIVE SKETCH OF A 3-SPACE ROOM INTERIOR THE PARALLEL-

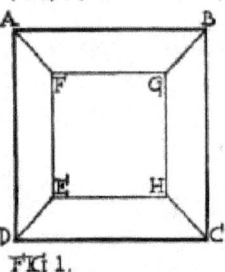

FIG 1.

OGRAM OF THE FURTHER WALL IS APPARENTLY CONTAINED WITHIN THAT OF THE NEARER. FIG 1 ILLUSTRATES THIS AND ALSO ANOTHER IMPORTANT POINT, VIZ. THAT THE APPARENTLY SLANTING LINES AND PLANES OF THE TESSERACT ARE PURELY AN ILLUSION OF PERSPECTIVE. IN FIG 1, A F, B G, ETC ARE PART OF A SYSTEM OF 3 SETS OF PERPENDICULARS, THOUGH THEY APPEAR TO SLANT, WHILE A B APPEARS PERPENDICULAR TO A D, ETC. FURTHER, OBSERVE THAT IT WAS THE EXIGENCIES OF REPRESENTATION THAT COMPELLED, IN THE PROJECTION, THE PLACING OF SPHERE 79 INSIDE OF 41, AND THAT INSIDE OF 03 — THEY REALLY FALL ONE BEHIND ANOTHER IN 4-SPACE.

IT IS INTERESTING AT THIS POINT TO IDENTIFY A FEW OF THE TESSERACT DIAGONALS AND TO VERIFY THEIR SUMMATIONS. JUST AS IN FIG. 1 A DIAGONAL EXTENDS FROM ONE CORNER OF THE OUTER SQUARE ACROSS THE CENTRAL, TO THE OPPOSITE CORNER OF THE INNER, SO IN THE TESSERACT WE PASS FROM THE OUTERMOST CUBE CORNER TO THE OPPOSITE CORNER OF THE INNERMOST CUBE, CROSSING AND INCLUDING THE TESSERACT CENTER, SPHERE 41 THUS: $-37+41+45=123$; $18+41+64=123$ ETC.

THE FOREGOING WILL SUFFICE TO EXPLAIN THE SKETCHES AND ENABLE THE READER TO EXAMINE THE CURIOUS ASSEMBLAGE OF CONCEPTS AND RELATIONS WHICH CONSTITUTE THE MAGIC TESSERACT.

The following research regarding the spiral coils for zero-point energy belongs to Prof Dr Konstantin Mel - scalar waves. From the extended Vortex and field theory to a technical, biological and historical use of longitude waves. Edition belonging to the lecture and seminar "Electromagnetic Environmental compatibility.

Space energy technology (SET)

16.9 Tesla's flat coil

In the category of the unconventional coiling techniques without doubt the Tesla coil may not be missing. If in schools and high schools such coils were standing in the laboratory for teaching purposes, then as a rule it are cylindrical coils. In reality Tesla worked with flat coils but that, so is said, isn't necessary anymore today, since we have at our disposal better isolating materials than 100 years ago. Actually Tesla contended with problems of isolation, which he could solve with the help of the flat coil, but it should turn out that the coil geometry is attached a crucial importance.

Everything had started with Tesla having to leave the Technical High school in Graz without diploma. He ran out of money and he had dared to criticize the venerable Professor Poeschel and his sparking Gramm dynamo. With that, he had put himself under compulsion to succeed. Two years later he had ready the solution. In the year 1882, he discovered the rotary field in Budapest.

In the time to come he designs and builds an alternating current motor, but no-one wants to have it and surely Thomas Alva Edison not. Tesla after this disagreement very fast gives up his job at the Edison Company again and again stands under pressure to succeed. With that the eternal bachelor Tesla urges himself to ever higher efforts. He wants to prove himself and the rest of the world that his alternating current system is superior to the direct current system. Direct current, as is well-known, can't be transformed, and thus the advantage of Teslas alternating current lies in the possibility of power transport by high-tension cable over large distances. But for that the high-tension transformers first had to be developed and thereby the said problems of isolation occurred. With each turn, the tension voltage at the transformer winding increases. The distance tothe grounding point lying on the outside has to be chosen bigger with each turn, so that no blow inside of the high-tension winding occurs. A consistent solution of the problem in accordance with engineering is the flat coil used by Tesla, wound spirally from the inside to the outside (fig. 16.9 A).

It thus is correct that isolation technical reasons led to the flat coil, since Tesla himself was completely surprised as he had to find out that this coil can lose its self-induction, that scalar waves can be detected with it and that it is cooled down during operation in an inexplicable manner. This cooling effect Tesla has investigated more detailed and after all even used. In his patent specification concerning the superconductivity he describes, that the flat coil also loses its Ohmic resistance, if he in addition previously cools it with liquid air. The remaining cooling down to absolute zero his flat coil obviously has carried out entirely by itself with help of the neutrinos (fig. 16.9 B)<ii>.

<i>: Nikola Tesla: Coil for Electro-Magnets; Patent No. 512,340 (1894)
<ii>: Nikola Tesla: Means for Increasing the Intensity of Electrical Oscillations,
Patent No. 685,012 (1901)
The secret of the flat coil

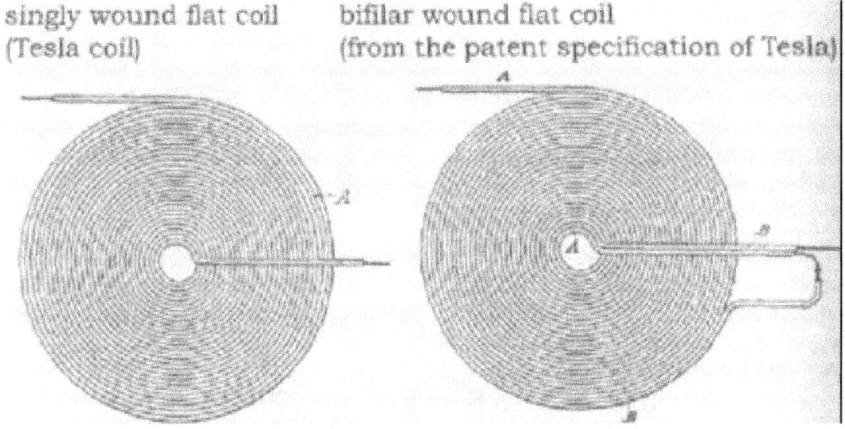

Fig. 16.9 A: Coiling techniques of Tesla's flat coil.<i>

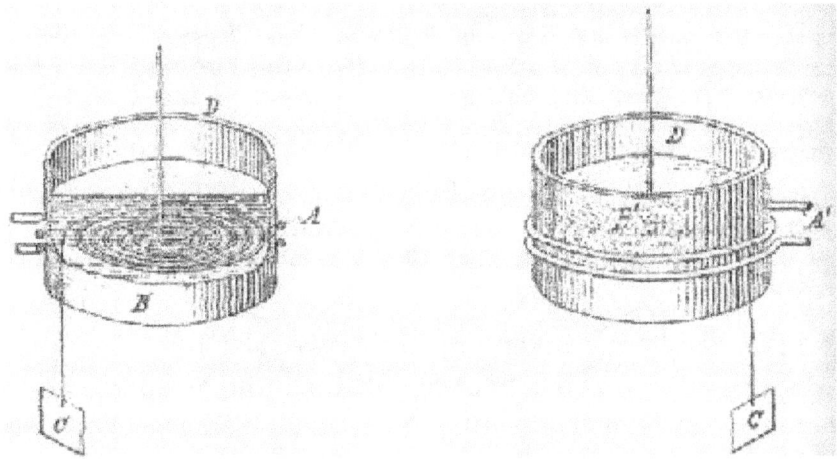

Fig. 16.9 B: Patent specification of Tesla concerning Superconductivity.<ii>
<i>: Nikola Tesla: Coil for Electro-Magnets; Patent No. 512,340 (1894)
<ii>: Nikola Tesla: Means for Increasing the Intensity of Electrical Oscillations,
Patent No. 685,012 (1901)
Prof Dr Konstantin Meyl mentions the measuring system in the section on the spiral coil he calls it a "strange Gauge"
Space energy technology (SET)

16.10 The secret of the flat coil

The technical function could be explained in the way that the charge carriers of a flat coil by induction are set into motion for excitation from the outside. The transmitted energy shows in form of kinetic energy. The spiral flat coil becomes narrower and narrower towards the inside, the length of each winding shorter and shorter, so that the kinetic energy inevitable has to decrease in favour of a rotational energy. The faster and faster rotating spherical vortices are pulled apart to flat discs and eventually to ring-like vortices by the centrifugal force. The electrons at first become neutrinos bound to a line and finally free neutrinos. Tesla has technically used the first ones in the single-wire-transmission technique (fig. 9.5) and the last ones in his wireless energy transmission (fig. 9.7).

Like many other inventors, Tesla owes also the inventions, which he counts his greatest, like the radio technique and the Magnifying Transmitter, first of all, his industriousness. His persistence and a great deal of inventor luck.

A magician, as he is called in his most important biography; he by no means was<i>. The flat coil, to which led him chance and which plays a central role in all these inventions, gave him the lucky position, to collect neutrinos and materialize them to charge carriers or in reversed direction to dematerialise electrons to neutrinos. The technology however is everything else but new. Already the Lituus of the Etruscan and Roman Augurs and the crook of the priests had the same spiral structure (fig. 16.10).

In the case of the devices, which the Augurs for instance served at land surveying, it clearly concerns flat coils according to Tesla. We will go into this strange "gauge" more in detail in part 3 of the book<ii>.
"The strange gauge is the Vedic metric"

The trick probably is, that one component of the electric field pointer is directed towards the centre of the coil and as a result, some open field lines are generated, which then collect neutrinos from space. In this process the neutrinos thanks to the resonant interaction are slowed down to the speed of light and following, as discussed, materialized by means of the flat coil, as in addition rotational energy is withdrawn from the neutrinos. Since the receiver oscillates resonant with opposite phase, in addition, the thermal oscillations are reduced and the receiver becomes cold!

If one compares the Mobius coil with the Tesla coil, then besides numerous properties in common the strength of the first coil lies in the production of open field lines and the collection of neutrinos, whereas the special and additional property of the flat coil lies in the materialization, in the conversion of neutrinos into charge carriers. However, the advantages of the flat coil have to be bought at the expense of having to work with very high tension voltages (above 511 kV) and with large changes in tension voltage (du/dt). With this set of difficulties, we will have to deal in more detail.

Margaret Cheney. Nikola Tesla, Erfinder, Magier, Prophet (Orig.: Man Out Of Time, 1981), Omega-Verlag Düsseldorf 1995 K. Meyl: Electromagnetic environmental compatibility, part 3, edition belonging to the information technical seminar, INDEL Verlagsabteilung 2003. 348 Discussion concerning the technology of the neutrino collectors.

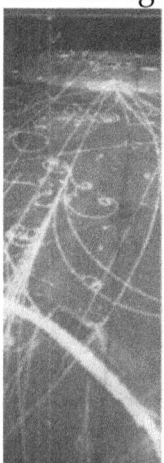

Fig. 16.10: The Lituus or crook of the Augurs in ancient Rome

The spiral Pictures and Quote are from Readers Digest "In to the Unknown" published, Sydney 1982.This pictures below were taken inside a bubble chamber, shows the delicate tracks made by bits of energy following a subatomic collision.

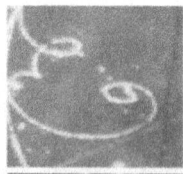

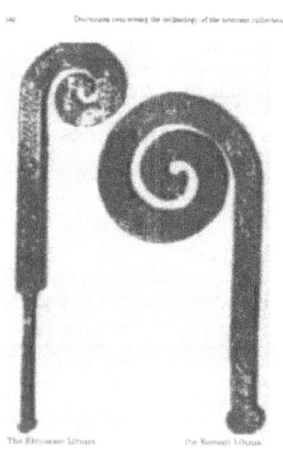

Fig. 16.10 The Lituus or crook of the Augurs in ancient Rome

Such constant, yet hard to imagine, exchanges challenge everyday concepts of time, space and reality.
The 2 coils above right are actual tuning coils for zero point energy wound using sacred geometry and the Vedic metric system. All of the proof is put together by Prof DR Konstantin Meyl President of the Society for the Advantage of Physics and Vice President of the German Association for Space Energy.

This shows that in ANCIENT ROME THEY USED ZERO POINT ENERGY
Space energy technology (SET)

16.11 Discussion concerning the technology of the neutrino collectors Let us again collect the facts for the discussion: A SET-device is distinguished by a more or less unipolar design and open field lines, with which interact neutrinos, which are oscillating in resonance.

These then are slowed down and collected. For the transient process a large change in tension voltage (dE/dt) is required, which can be obtained directly, for instance by means of a spark gap, or indirectly by means of Faraday's law concerning the unipolar induction ($E = v \times B$).

The discussed possibilities concern the acceleration of a machine part (dv/dt), the variation of the magnetic field (dB/dt) by pulse-like excitation signals (16.5) or by magnetic flux variation (16.3) and the rail gun. Which even can be operated without foreign magnetic field (fig. 15.5 C), for which in that case occur both a dv/dt, and at the same time a dB/dt.

For resting arrangements, the velocity v is that of the charge carriers moving in the conductor. So that Faraday's law thereby does not lose its influence, the pointers of E and v must not point in the same direction, as in the case of "normal" coils. Unconventional windings, which for instance can be knotted like Mobius strips (16.7), take remedial action. Also the ancient crook, rediscovered as flat coil of Tesla (fig. 16.10), proves to be suitable in principle. Here one component of the electric field pointer points in the direction of the centre so that the wanted, at least partly, unipolar arrangement can be formed.

The first step, the collecting of space quanta, should not pose an insurmountable obstacle anymore in view of the numerous possibilities and the detailed explanations. A real difficulty we still have before us, because in most cases some ring-like vortices bound to a line are formed, for which no electronic construction element exists and for which functioning converters hardly are known.

There spoons are bending, some lumps are flying through space, radioactivity is disappearing without a trace, light phenomena are formed and the device suddenly is becoming cold. Almost all inventors, who have arrived in this place, are enthusiastic about the not understood effects or with that are wanting to get attention, but hardly anyone really starts something with that. Until now, the necessary system and a useful theory were missing.

Only too often isn't considered, that only an indirect conversion into charge carriers is possible, that during the materialization of neutrinos an intermediate product is formed, which can be described with the model concept of a neutrino bound to a line or of an oscillating ring-like vortex. The technologies collected in this chapter concerning the collecting of neutrinos only form the first step from the free to the bound ring-like vortex.

Cathode Light-Multiplying Compression Rings

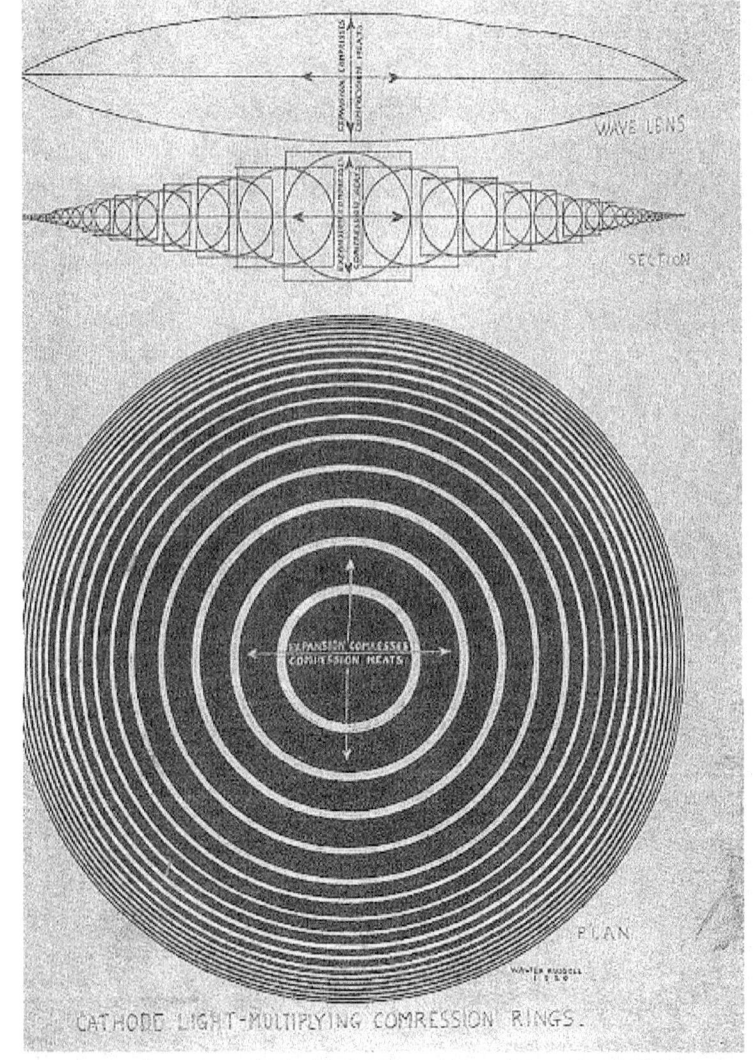

Completed in 1960 and published in *The Home Study Course* XI: 795.

The above image by Walter Russell from "In the wave lies the Secret of creation" and I will now show you how to tune into natures energy using spiral flat coils.

See Fig. 16.9 B: Patent specification of Tesla concerning Superconductivity Page 150 to see these coils. I am giving away some of this information as to show people what can be achieved if you believe in yourself and don't follow the heard like a sheep. There are so many people out there trying to do the right thing but don't trust their own instinct.

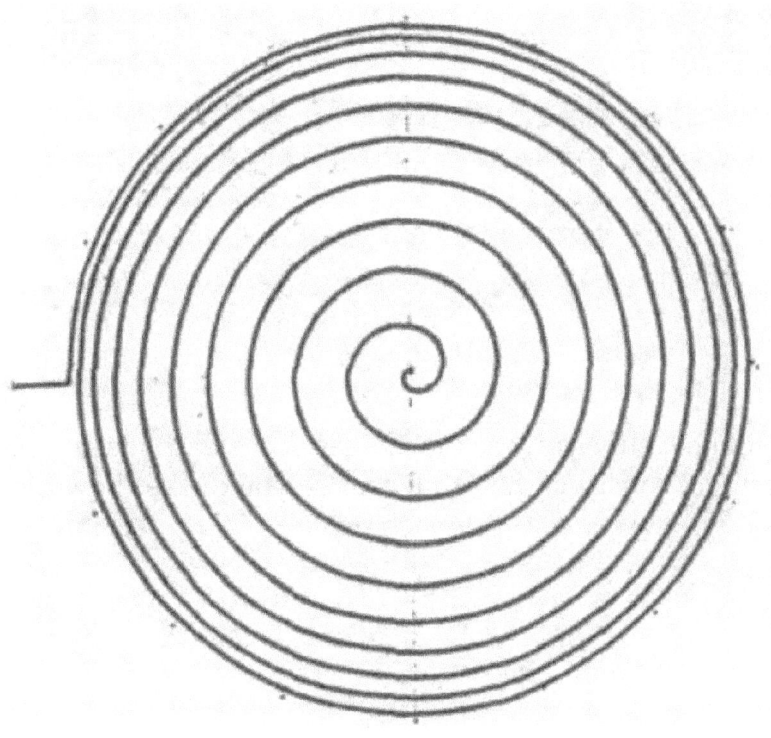

This Image of a coil is from "Pentagon Aliens" by William Lyne and below is the circles to construct this spiral.

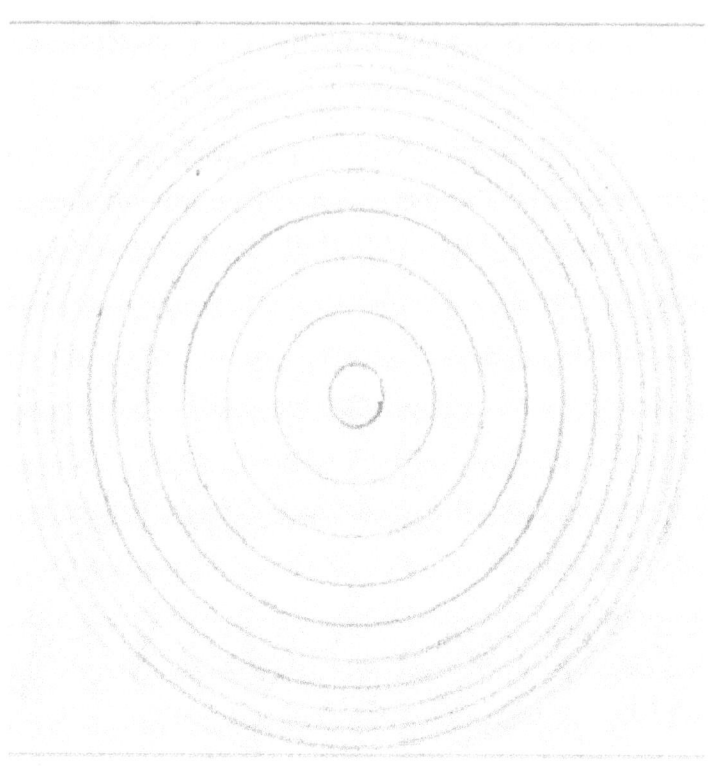

I will now superimpose the images of the Cathode and the circles to construct the spiral coils so you can see how they match up.

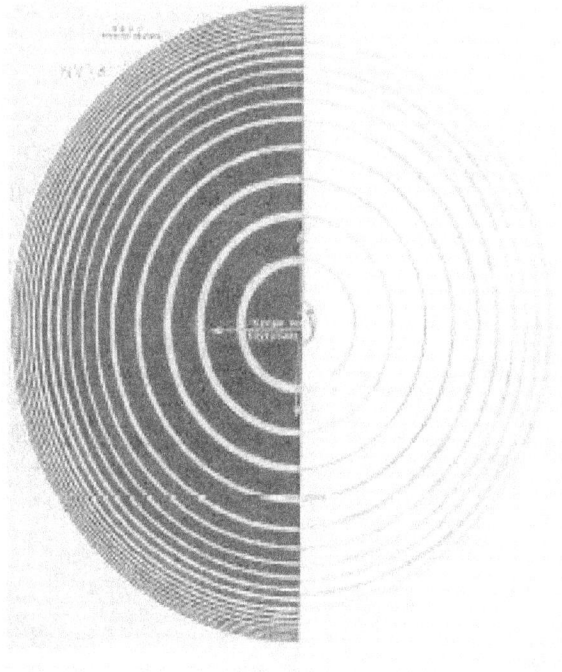

The below image is the overlay of all three images. This is the foundation of the tuning coil to tap into nature's energy. Or hacking the matrix. This is how Tesla and many others throughout history have used Ether Physics.

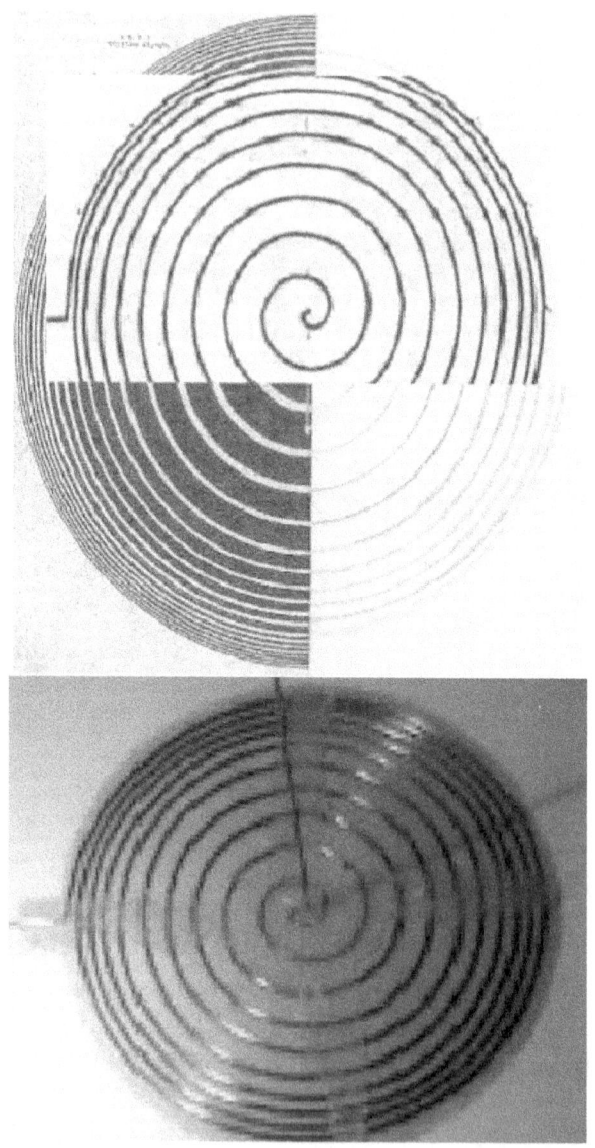

These coils can be pulsed using HF negative DC power supply. The coil is represented in spiral format and this is how you create the coils. This coil is XX.Xcm in diameter and is XXXXmm long "exactly" use a HV negative charge to give the coil an initial charge then it acts like a capacitor and keeps charging itself.

Do not earth it out because that short-circuits it "open circuit only" You will get a strong smell of ozone when the coil is working correctly. The coil vibrates when in use AND THEN after working for a short period of time it reduces in temperature and the coil approaches zero degrees, this is where the term zero point comes from.

These coils are being developed as a standalone system in Auckland New Zealand and the exact tuned length will not be disclosed. This was just one format tried but not the end product.

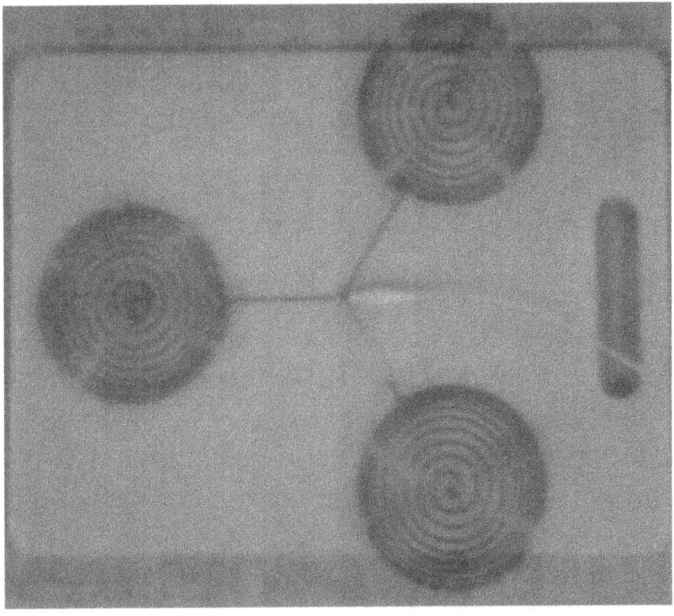

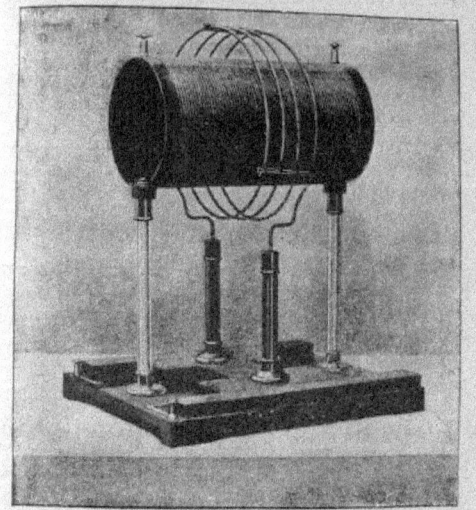

High-tension Coil," (See Fig. 7), which is in reality a High-frequency Transformer of the *Tesla-Thomson* type, air being used for insulating purposes instead of oil. At the present time these devices are seldom used among European specialists, their High-potential, High-frequency Currents being generally derived from what is known as "*Oudin's* Resonator"; (See Fig. 9).

FIG. 7.—D'Arsonval Bipolar High-tension Coil. (Williams.)

This device consists of a large fiber cylinder or tube, having a number of turns of fine insulated wire wound on a spiral upon its outer surface. The lower end of this wire is connected to the Small Solenoid of *d'Arsonval*. As a rule the latter is incorporated in the resonator and consists of fifteen or twenty turns of course copper wire wound upon the lower part of the cylinder, as in Fig. 9, the upper end of the solenoid being continuous with the lower end of the Resonator Coil. (See Fig. 8). It is necessary that a certain inductive relation exist between the solenoid

The four more stout coils of wire wound around the 100 small gauge windings are what make this coil interesting. The four thick gauge windings are wound at the circumference at the centre of the wire to 360 mm circumference so that is 4 x 360 = 1440 mm and the 100 windings are at 216mm circumference because of the 360/216 = 1.666 perfect harmonic balance, so 100 turns x 216mm = 21600

Everything must be tuned in and all lengths of wire must be measured exactly.

heightens its tendency to pass into the primary coil at the ends, where it must be therefore specially insulated from it.

In winding these sections there is a

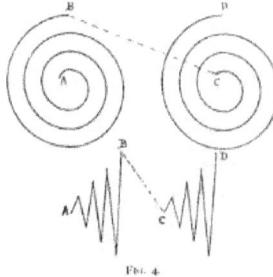

Fig. 4.

method now generally adopted which has many good points, although at first it may seem complicated. The old way of filling two sections was to wind both in the same direction as full as desired, then join the outside end of the left-hand coil to the inside end of the right hand coil. This necessitated bringing the outside end down between two disks, or in a vertical

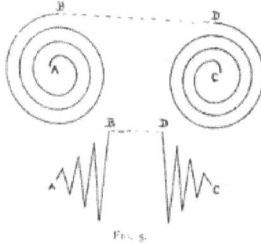

Fig. 5.

hole in the sectional divider, and thereby rendered it liable to spark through into its own coil. This is shown in Fig. 4. A and C inside ends, B and D outside ends, the disk being between B and C.

Fig. 102.

the properties of a separate magnet. Fig. 102 is intended to give an idea of the mode of arrangement of these elementary magnets; they are, of course, so minute as to be beyond the reach of the most powerful microscopes. All their poles lie in the same direction, as indicated by the light and shaded portions in the cut; the light half of each little magnet representing its north pole, and the shaded half its south pole. These poles are distributed through the whole length of the bar; but in consequence of their mutual reactions, the resultant polarities are strongest near the ends of the bar, becoming more and more completely neutralized towards the middle.

The above image is just another example of tuning coil which were used more than 150 years ago. You have been shown the Cathode coil and therefore there should be an anode coil too. I thought I would throw this one in to show that magnets are not just North and South Pole but the pole are divided throughout the magnet.

AC motor:
I have Reverse Engineering a modern electric motor and water pump that uses the correct thinking to obtain an excellent level of efficiency. I will also explain how they could have made it better. The motor and pump are made in Italy and retail for around two grand but read my reverse engineering of it will help you If you are struggling to come to terms with this information I will endeavour to explain it more clearly for you with this off the shelf product.. (This pump and motor were in perfect working order before I destroyed it to give you this example.)
The Alternating Current Motor is 90 mm in length and has

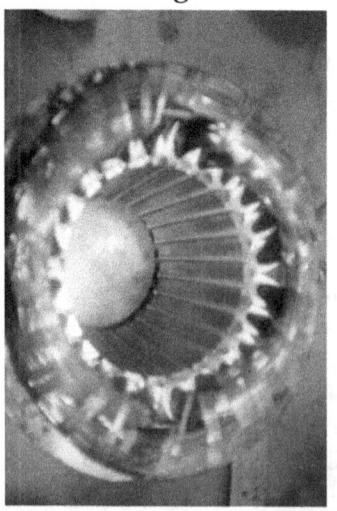

24 segments upon the interior of the tin plate core of the cylinder as it contains 8 sections of 0 .6 mm copper winding wire and 16 sections of 1.0 mm copper winding wire. 8 + 16 =24 sections in total.

1) First of all the total average length of each complete loop is (45mm x 2) + (90mm x 2) = 270 mm or 1 Vedic metric foot. The 4 sections (8/2 as a loop covers 2 sections) have 48 turns of 0.6 mm wire per section at 270 mm per section of the loop. 270 mm x 48 turns in each section x 4 sections = 51840 mm. **51840 mm / 24 mm inch of the 1080 Vedic metric = 2160 inches (sine value of 0)**

2) The 8 sections (16/2 as a loop covers 2 sections) have 48 turns of 1.0 mm wire per section at 270mm per section of the loop. 270 mm x 48 turns in each section x 8 sections = 103680 mm

103680 mm / 24 mm inch of the 1080 Vedic metric = 4320 inches (sine value of 0)

3) So the total length of wire used for this motor is (51840 mm + 103680 mm) = 155520 mm (155520 / 24 mm inch) = 6480 inches or (155520mm / 1080mm Vedic metre) = 144 meters.

(2160 inches + 6480 inches) = 8640 inches 25920 / 8640 = 3, one third of 25920 elliptical wobble of the earth which is what The Coral Castle device of Ed's achieves.

The core is the key to making it work as they retard or lessen the core by 10 degrees out of phase from our perceived reality and all so the materials used should be soft iron not multiple sheets of poor quality pressed tin shacked together. The motor and winding is 138.20 x by π = 434.16 just out of harmonic range of 432.

The thickness of the wire is also a key point as the length of the wire is 8640 and the two-length thickness of wire are 1 mm and 0.6 mm. 8640 / 0.6mm = 14400 the thickness of the wire must be a harmonic the length of the wire, everything is harmonics. The actual core is 54.13 mm in diameter x π = 170.05 mm. They should have made it to 57.29577951 mm x π = 180 perfect zero sine wave. If you have, a calculator with a SINE function put in 180.. This would make that motor far more efficient using a lot less power; industrial has governed itself to intentionally deceiving people into using more energy than necessary. "It is A Sort or price fixing if I can put it that way".

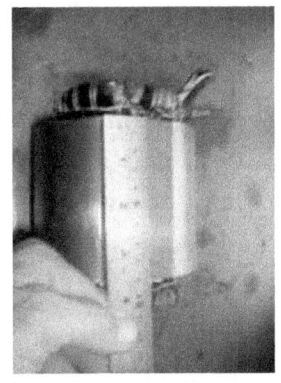
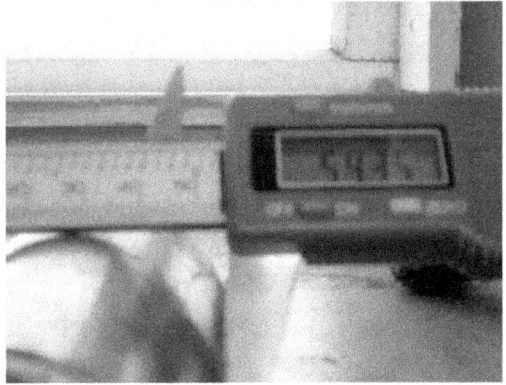

The Pump.

The diameter of the water pump high-pressure outlet is 137.5098708 mm and that multiplied by π (3.14) = 432 and if you look below that is musical Note A.

That part the harmonics for the water pump, as you will see there are five outlets for the water to exit in one 360-degree motion 360 / 5 =72. The reason this does not run in perfect harmonic balance and has a balancing groove cut into it is the poor manufacturing and the steel rotor section of the pump it is thicker on one side than the other is, and should be 432 diameters. Where the rotor of the pump sits in the housing is exactly 144 mm from the back to give the 144 by 432 relationship. The overall length of the housing is 16 cm.

This is another example to help you. This is an old motor I got from a market for $5. The outer of the motor is 135.03 mm in diameter that is 424.21 mm circumference, if made correctly it would be 432 exactly. The core is 70.63 mm in diameter when they should have used 72 mm for 72 beats per min 432/72=6. The width of the core is 37.16 when it should be 36 mm. When I cut the wire out of the burned out section and counted how many turns they did I got the sum of 547 and at the average length pictured below of 19 cm 10393 then divide that by 24 mm inch you get 433.04 it should be 432. There are four windings at 432 and that is 1728. Then if you divide 1728 by 24 = 72. There are two different features on this old motor, it has a vortex fan that is built into the core of the motor also this motor has copper at either end of the motors core. 1728 also has to do with the calculation of the mm value.

The seal of Solomon or Star of David which ever you prefer to use has to do with the flow of energy whether it be within the human body our planet or even electrical motors.

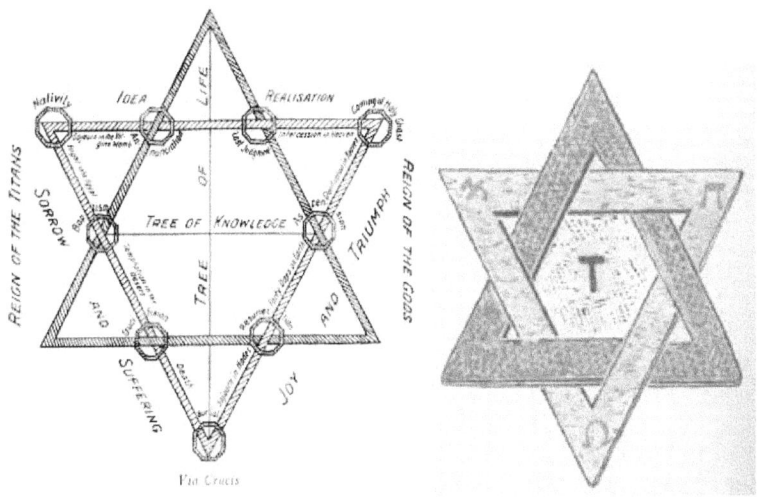

The images below are from "Theory and Calculation of ALTERNATING CURRENT PHENOMENA" BY CHARLES PROTEUS STEINMETZ "3RD EDITION 1900.

TRANSFORMATION OF POLYPHASE SYSTEMS. 465

5. The L connection for transformation between quarter-phase and three-phase as described in the instance, paragraph 257.

6. The T connection of transformation between quarter-phase and three-phase, as shown in Fig. 203. The quarter-phase side of the transformers contains two equal and

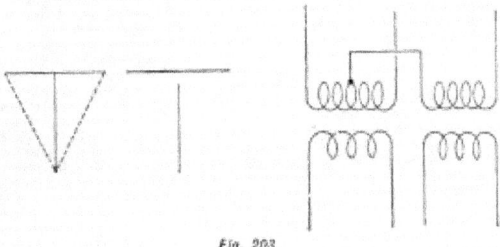

Fig. 203.

independent (or interlinked) coils, the three-phase side two coils with the ratio of turns $1 + \dfrac{\sqrt{3}}{2}$ connected in T.

7. The double delta connection of transformation from three-phase to six-phase, shown in Fig. 204. Three transformers, with two secondary coils each, are used, one set of

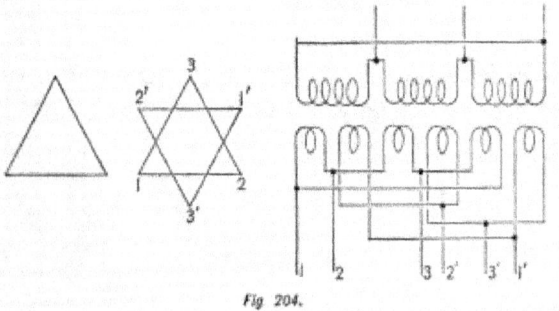

Fig. 204.

secondary coils connected in delta, the other set in delta also, but with reversed terminals, so as to give a reversed E.M.F. triangle. These E.M.F.'s thus give topographically a six-cornered star.

466   ALTERNATING-CURRENT PHENOMENA.

8. The double Y connection of transformation from three-phase to six-phase, shown in Fig. 205. It is analogous to (7), the delta connection merely being replaced by the Y connection. The neutrals of the two Y's may be connected together and to an external neutral if desired.

9. The double T connection of transformation from

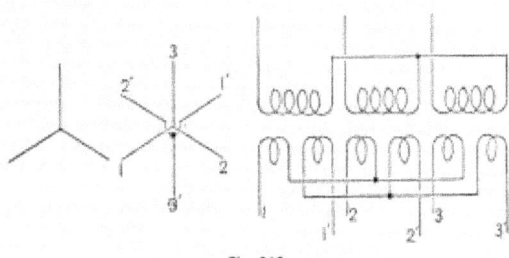

Fig. 205.

three-phase to six-phase, shown in Fig. 206. Two transformers are used with two secondary coils which are T connected, but one with reversed terminals. This method allows a secondary neutral also to be brought out.

287. Transformation with a change of the balance factor of the system is possible only by means of apparatus

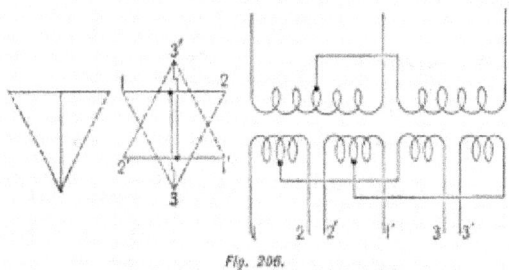

Fig. 206.

able to store energy, since the difference of power between primary and secondary circuit has to be stored at the time when the secondary power is below the primary, and returned during the time when the primary power is below

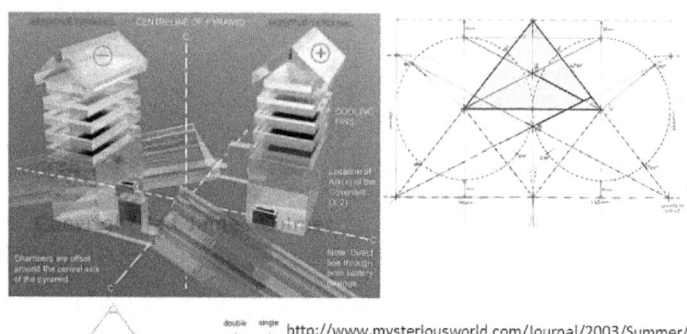

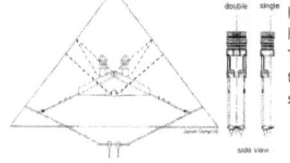

http://www.mysteriousworld.com/Journal/2003/Summer/Osiria/
http://sentinelkennels.com/Research_Article_V41.html
The pyramid pump and the Gaza power plant hydrogen reactor explains what the pyramids were designed to do. You can see they have blocked off section of the pyramids and are hiding something.

Pyramid power. Pictures of the missing inner chambers by James Colmer.

The following research regarding Scalar wave technology in antiquity belongs to Prof Dr Konstantin Mel - scalar waves. From the extended Vortex and field theory to a technical, biological and historical use of longitude waves. Edition belonging to the lecture and seminar "Electromagnetic Environmental compatibility.

Scalar wave technology in antiquity

The end of the book about potential vortices and their propagation as a scalar wave shall form an impressive example, where as many of the derived wave aspects as possible have an effect. It shall be proven that already in antiquity radio engineering based on scalar waves has been used. The proof starts with a thesis.

30.1 Thesis:

The temples in antiquity all were short wave broadcasting stations.

And energy from the field served as an energy source, so e.g. the earth radiation in the case of temples of terrestrial gods. In the case of the solar god the radiation of the sun was used, whereas for the temples, which were dedicated to the planetary gods, the neutrino radiation arriving from the planets served as an energy source. If the temple was dedicated to a particular god, then the name of the god was representing the used frequency of the broadcasting company. The corresponding wavelength, resp. the respective god, understandably was "immortal".

Not so the broadcasting technicians on duty, who as human beings naturally were mortal, who took turns in the studio as members of the priest council and who merely had to impersonate the god Apollo, Poseidon etc. by the name of the broadcasting company, if they went on air. Only for the news editor Homer and for few of his colleagues we actually know the names of the persons behind the scenes. In the temple books the texts have been recorded, which a god and its broadcasting company have received. The chosen meter served the easier detection and correction of transmission errors.

Here often a lot of fantasy was necessary, for which reason the reception facilities commonly were described as oracle<i>. The reception of the news as a rule took place on an altar. Thereby the direct effect of scalar waves on man e.g. in the case of the so-called temple sleep or the indirect influence on biological systems, e.g. on the intestines of slaughtered animals, was evaluated.

A further development of the telegraphy was the tripod technology<ii>, in which case by turning of the polarization plane individual symbols and letters were transmitted up to the transmission of the spoken word with the help of a special wavelength modulation.
That far the thesis reaches, which now should be proven.

Prerequisites.

```
godology = high frequency technology
          God name = RDS, station identification
          Members of a = broadcasting studios of a
          Family of gods broadcasting company
          Pantheon, = extremely broadband
          Temple of all gods FM broadcast station
          Crown = antenna netting
          Gifts for sacrificing = broadcasting fees
   Place of sacrificing = place of a node of the standing wave
          Earth radiation = power supply Homage of a = time
          restriction of the Weekday operation of the station
          Zeus , forges = electrostatic blows, when Thunderbolts a
                    temple is oscillating
          Ritual act = technical provision for transmission and
                              Reception
          Pythia of Delphi = radio telephone operator, receptionist
          Cella (marrow of temple) = tuned cavity
                    Obelisk = antenna rod
```

Fig. 30.2: ____ little dictionary for ancient radio engineering (2).

30.2 Prerequisites

The argumentation has to be made on mathematical-physical foundation. The prerequisite for that are the 29 chapters of before. The following points could be demonstrated and derived:

1. The wave equation (inhomogeneous Laplace equation) describes the sum of two wave parts, where

2. Every antenna emits both parts,

3. The transverse part, known as electromagnetic wave (Hertzian wave)

4. And the longitudinal part (Tesla radiation) termed scalar wave by the discoverer, better known as antenna noise.

5. The wave equation mathematically describes the connection of both wave parts in general and the conversion of one part into the other in particular, thus

6. The rolling up and unrolling of waves in field vortices (measurable as noise).

7. The transition takes place proportionally to the Golden Proportion, as resulted from the derivation (chapter 29.7 - 29.9)

With the last point the electro technical problem becomes a geometrical problem, if it concerns the use of scalar waves. The geometry of the antenna is crucial. Thereby the Golden Proportion provides the necessary direction for construction.

That justifies the assumption that the buildings in antiquity, which were built according to the Golden Proportion, were technical facilities for scalar waves. Maybe the builders had specifications that had physical reasons and could mathematically be proven.

At this place there result completely new aspects for judging and interpreting buildings especially from antiquity through the derivation of the Golden Proportion from the fundamental field equation. If we have understood their way of functioning, then we will be able to learn much from that for our own future and for the future construction of scalar wave devices.

As a further prerequisite for the ancient broadcasting technology enough field energy should be at disposal. We proceed from the assumption that:

1. The earth magnetism and the cosmic neutrino radiation are tightly hanging together by the processes in earth's core,

2. The earth magnetism in antiquity verifiably was approx. thousand fold stronger than today (proven by gauging of pieces of broken pot),

3. As a consequence the neutrino radiation in antiquity as well must have been thousand fold stronger and

4. The cosmic neutrino radiation has served the transmitting plants of antiquity as an energy carrier; any thought is absurd to reject the technical function of a temple only because it today can't be reproduced anymore. The artistic and aesthetical viewpoints, which are put into the foreground by art historians because of ignorance about the true function, rather are secondary.

The terms used to describe the broadcasting technology in antiquity in the last 2000 years have experienced a shift of meaning, so that a translation in our linguistic usage of today is necessary. The adjacent dictionary should help in that case.

Not everyone, somehow participating in send receive engineering, at the same time also was inaugurated in the entire secret knowledge. Most priests only knew as much as they necessarily needed to know to fulfill their tasks. Thus a temple priest, who was resented an enciphered text and who should bring this on the air, not necessarily at the same time needed to know the content of the text or the code. The same of course also was valid for the sacrificing priest acting in the receiving station. The Vestal virgins for instance had to present the received text to the Augures, by whom they were supervised and controlled.

But who wanted to introduce a new god in the gods heaven and perhaps even himself be worshipped as a god, should have complete command of both the broadcasting technique and the reception technique. In ancient Egypt the Pharao at least once a year had to prove, that he still was in command of the technique. Otherwise he was replaced. For a person with security clearance that at the same time was a death sentence.

In the historical facts numerous pieces of circumstantial evidence can be found, which can be considered to be evidence for the thesis of the operation of send receive engineering in antiquity. One now perhaps understands, why the rulers were put an antenna netting over their head, a so-called crown, or why the Augures could survey the land with a flat Tesla coil in their hands (fig. 16.10).

Direct evidence is present as well. It can be found in ancient texts. But it is questionable if historical texts concerning ancient radio engineering have been translated correctly. The talk is about oracles, mystery cult and earth prophesy if the receiver is meant.

The predominantly technically uneducated historians attest the Romans a defective sense of time, because their couriers surely could not cover the long ways across the Roman empire so fast at all, if they read in the Latin texts: "They sent by courier to the emperor in Rome and got for answer...". The answer of the emperor namely already arrived at the squad at the latest in the following night. The correct translation should read: "they cabled" or "they broadcasted to the emperor in Rome and got for answer. Such a big empire as the Roman Empire actually only could be reined by means of an efficient communication. Cicero coined the word: "We have conquered the peoples of the earth owing to our broadcasting technology...<iii>"! The term broadcasting technology from ignorance is translated with piety. If engineers however rework the incorrect translations, then one will discover that numerous texts tell of the broadcasting technology, that thus correspondingly much direct evidence exists concerning the practical use of this technology.

For the Roman military transmitters, which formed the backbone of the administration of the empire, the reading off of the information from observations of nature like the bird flight or from felt signals of a geomanter was too unreliable.

They read off the information from the rhythm of the convulsions of the intestines of freshly slaughtered animals. In the case of the dead animals on the altar every extrinsic influence was excluded. But the enormous need of slaughter cattle was a disadvantage. Who wanted to have information, first of all had to bring along an animal, which then was, sacrificed" the god, or better say, which was abused as a receiver for a particular transmitter. Thereby the innards served as a biosensor and as a receiver for the news.

30.5 Radio technical concepts
In planning and constructing radio technical networks only a few possible concepts exist.
It is interesting that at least one historical example can be specified for every concept. That shows that all possibilities were tried at least once. The three most important concepts are presented here:

A. Cellular phone.

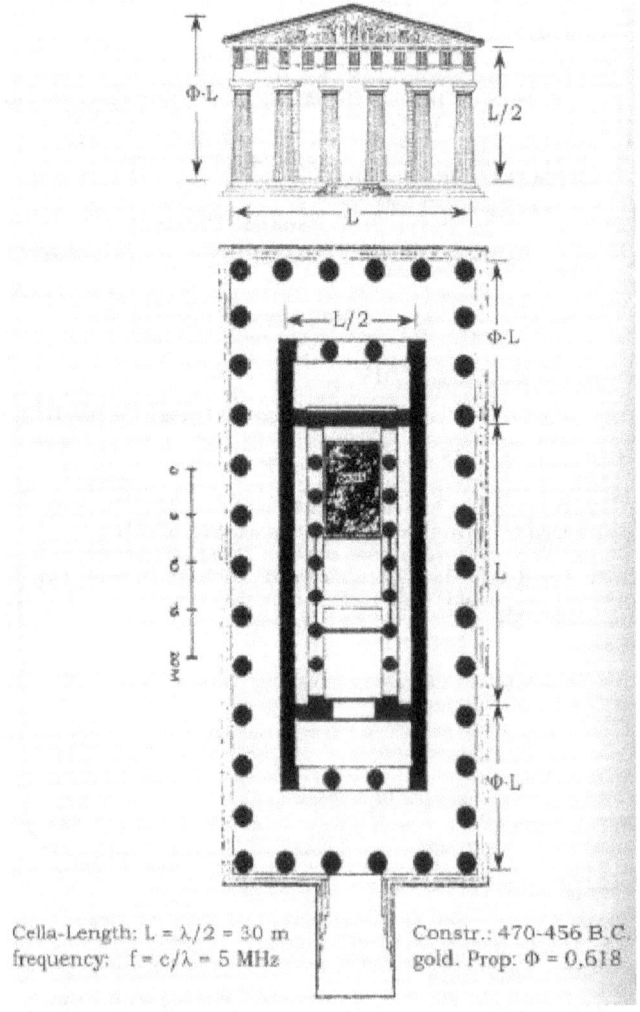

Fig. 30.3: The Golden Proportion of Zeus-temple in Olympia.[i]

[i]: K. Schefold: Die Griechen und ihre Nachbarn, Propylaen Kunstgeschichte Berlin Bd. 1, Abbildungen von Seite 249

Every bigger city between Euphrates and Tigris, which thought the world of itself, had at its disposal already in antiquity a temple tower. Such a temple tower was a, telephone cabin" in the form of a pyramid as a transmitter and a receiver temple at the top, to where the receptionist adjourned to the so-called temple sleep. Discipline was required, since all the time only one priest was allowed to broadcast. All others could listen to him doing so. If he was ready, he closed his contribution with a fixed symbol or term (over) and the next one could continue. This is a classic link-up, where anyone communicates with every network participant.

The stations all were strikingly similar in form and size of building, like one phone box resembles another. In that way a further development of the cellular phone system hardly was possible and that has a technical reason, as the building of a tower in Babylon has shown us. This tower namely had gotten the ambitious builders too big, so that the frequency of the Mesopotamian radio network had been left and instead a foreign network could be received, the code of which no-one could understand. The result was a confusion of language and the order to stop the building.

B. Broadcasting.

Millions of TV spectators every evening look in the ABC news or another daily journal of a TV Channel. In the case of broadcasting thus many receivers listen to the news of a powerful transmitter.

With that the whole plenitude of power is concentrated in the hands of the chief intendant. In antiquity he called himself high priest. If he went on the air, he used the logo of the god that he had to represent. Today the logo of the broadcasting company is shown in a corner of the TV screen. Even this very day feedback from the receiver to the transmitter hardly is possible contingent on principle.

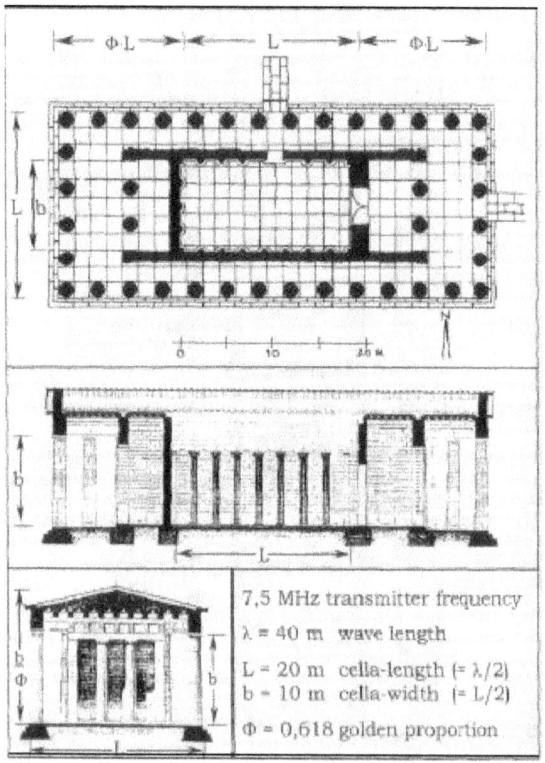

7,5 MHz transmitter frequency
λ = 40 m wave length
L = 20 m cella-length (= $\lambda/2$)
b = 10 m cella-width (= L/2)
Φ = 0,618 golden proportion

Fig. 30.4: Example Tegea, temple of Athena Alea.[i]
Built 350/340 B.C.

---

[i]: G. Gruben: Die Tempel der Griechen, Wissenschaftliche Buchgesellschaft Darmstadt 1986, 4. Aufl. Seite 130
[ii]: E. Horst: Konstantin der Grosse, Eine Biographie, Classen Verlag 1985 Dusseldorf, 2.Aufl., S. 89.
[iii]: E. Horst: Konstantin der Grosse, Eine Biographie, Classen Verlag 1985 Dusseldorf, 2.Aufl., S. 33.

The problems with nationally controlled broadcasting, with politics controlled by the media all are not new. The monotheism in ancient Egypt with the claim of lordship of the main god Ammun Re is an example from antiquity.

C. Dispatch service.
In ancient Greece the technical structures and with that also the power structures had been turned around. At that time a big network of broadcast stations, which continually was extended by a policy of settlement ordered by the gods, supplied a central and correspondingly powerful agency with information per radio. Who wanted up to date news, could call for these in the agency with seat in Delphi, but he had to pay for it. To accommodate the broadcasting fees in form of gold and gifts whole treasury stores had to be built. Measured by the commercial success the ancient news network has remained unmatched, and can't be compared with pay-TV or todays dispatch services, like dpa. If the network however becomes too big, uncontrollable and it lacks discipline, then it sometime will crack and the system crashes.

30.6 Wireless telegraphy Radio engineering 100 years ago also started with telegraphy. Thereby the high frequency carrier is switched on and off. With this technique Marconi succeeded in a radio transmission over the English Channel (1899) and over the Atlantic Ocean (1901). As next step the amplitude modulation (AM) followed. Thereby the HF-carrier is overlapped with the low-frequency signal of a sound carrier in such a way, that the amplitude fluctuates on the beat of the LF-signal.

As a disadvantageous effect, also noise signals will overlap, from which the quality of reception will suffer. Only the frequency modulation (FM), where the LF-signal is transmitted as temporal fluctuation of the frequency, brings an improvement.

The annoying amplitude noise hence has no effect in the case of FM. It easily can be recognized, how the development of the modulation techniques follows the urge for technical improvement and optimization. That in antiquity hasn't been different, for which reason the progress of development took place in the same order. The broadcasting technology of the ancient gods started with the wireless telegraphy.

Cellular phone.	
• *Network:*	All „telephone cabins" are transmitter and receiver at the same time.
• *Advantage:*	Anyone can communicate with every network participant.
• *Characteristic:*	The stations are strikingly similar in form and size of building.
• *Power structure:*	All stations as a rule have equal rights among each other.
• *Disadvantage:*	Innovations hardly are possible. The system is inflexible.
• *Examples:*	Temple towers (Iraq); cellular phone network.
Broadcasting.	
• *Network:*	Many receivers listen to the news of one powerful transmitter.
• *Advantage:*	The system is very flexible and permanently experiences improvements.
• *Characteristic:*	Monotheism. Big diversity among the receiver constructions.
• *Power structure:*	Central power concentrated in the hands of the chief intendant.
• *Disadvantage:*	Feedback from receiver to transmitter is not desired.
• *Examples:*	God Ammun Re (Egypt); ABC daily news.
Dispatch service.	
• *Network:*	Network of broadcast stations supplies an agency (receiver) with information.
• *Advantage:*	Optimal financing, since information is given away only for broadcasting fees in cash.
• *Characteristic:*	Treasury stores have to be built because of the immense riches.
• *Power structure:*	The power is concentrated in the hands of the receiver-agency.
• *Disadvantage:*	In the case of overcharge or overload a system crash is impending.
• *Examples:*	Pythia (Delphi); German Press Agency (dpa).

Fig. 30.5: _____ Three radio technical network structures, with an example from antiquity and from present time.

This is expressed in the architecture. Since electric resonant circuits or other frequency determining equipment weren't at the disposal of the engineers in antiquity, the determination and allocation of the broadcasting channels had to take place by means of the wavelength. The formation of a standing wave in the Cella, the innermost sanctuary of a temple, occurs if its length corresponds to half the wavelength of the HF-carrier.

The Roman architect Vitruvius calls the wavelength the„ basic measure", from which results "the system of the symmetries". He writes: The design of the temples bases on symmetry, to which laws the architects should adhere meticulously. "<ii>. „The length of the temple is partitioned in such a way that the width is equal to half the length, the Cella itself including the wall, which contains the door, is one fourth longer than wide. The remaining three fourths, which form the Pronaon, should protrude until the antae of the wall and the antae should have the thickness of the pillars"<iii>. If we recalculate ourselves, then the partitioning in 3/4 to 5/4 produces a proportion, which conies quite close to the Golden Proportion. In building a temple nothing is left to chance, after all it concerns the construction of a tuned cavity, capable of self-resonant oscillations with favourable emission behaviour.

From the outside one can't see if a telegraphy transmitter has been changed over to speech transmission with AM. The HF-carrier merely isn't switched off anymore, i.e. the priests let the temple oscillate without interruption. Newly added for AM is an electro acoustic coupling. For that many temples were retrofitted with a mouthpiece. Newly built AM transmitter temples conclude the Cella with a round apse.

Because of this acoustically conditioned construction the Cella length didn't have a fixed value anymore and the transmission frequency had become variable. Measured in the middle of the apse the wavelength was larger than at the sides, so that on the beat of the spoken word not only the amplitude of the field distribution in the interior of the temple, but in addition also the frequency of the self-resonant oscillation was changed. A typical example of such an architectonic hybrid form of AM and FM is situated in Rome. Because due to the frequency variation more than only one wave band was occupied and the temple consistently carries the names of two deities. It is the temple of Venus and Roma.

30.7 AM temple technology

The low-frequency signal (LF), which should be transmitted by a transmitter with amplitude modulation, lies in the range between 16 Hz and 16 kHz. If it only concerns the transmission of speech information, then the bandwidth can be reduced to 300 to 3000 Hz. In the case of mixing the low-frequency useful signal with the HF-carrier, thus in the case of the modulation of the carrier in the rhythm of the LF, two side bands arise.

These lie close to the carrier frequency and are formed from this once by the addition and once by the subtraction with the frequency of the LF-signal. Let's take the temple of Venus and Roma with a transmission frequency of 6.8 MHz If sound of 3 kHz should be transmitted clearly understandable, then the dimensions of the Cella had to be varied for just 8 mm for a corresponding Cella length of = 22 m.

As a curiosity the niches in the side walls in the case of this temple however allow a considerably larger bandwidth of more than 10% instead of the necessary 0.04% in the case of AM. In the case of the Greek originals, the Cella however has smooth walls, from which follows that the temples were designed ideally narrow band.
The Greeks apparently operated predominantly telegraphy transmitters, for which the side bands coincide with the carrier.

The argumentation indeed has remained unchanged: The modulator being narrow band and simple to realize speak in favour of the telegraphy being the "original form" of all modulation techniques. Also the rediscovery of the broadcasting technology by Heinrich Hertz succeeded as telegraphy signal. In addition the range is bigger than for any signal modulated with sound frequency. As the calculation example has shown, also pure AM transmitters work very narrow band, and this is particularly important for low transmission frequencies, if many transmitters want to use the favoured SW band between 3 and 10 MHz at the same time.

With AM one thus accommodates the maximum number of broadcasting channels in a particular frequency range, for instance the 80-meter band, without these interfering with each other too much. But that also was badly needed.

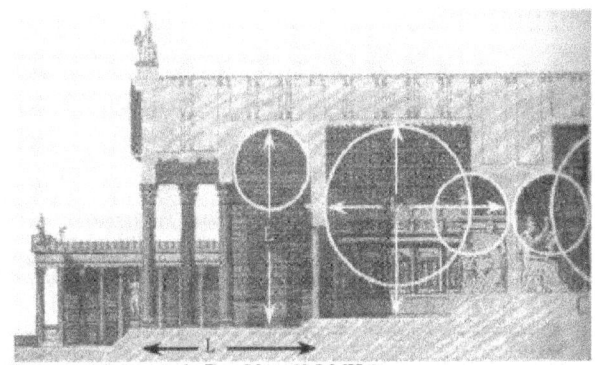

Diameter bigger circle $D_1 = 22$ m (6.8 MHz), small circle $D_2 = 11$ m; und $L = D_1 + 1/2\, D_2 = 27.5$ m (5.5 MHz)

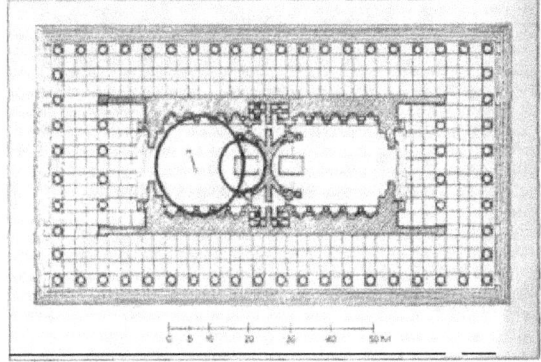

Fig. 30.7: Temple of Venus and Roma; Rome 136/37 A.D.[i,ii]

Conclusions about the everyday life of broadcasting in antiquity by all means are possible because of the enormous number of temple installations, which logically were permanently used. Only in Rome there existed up to 200 temples! Who goes in search of broadcasting stations with a modern short wave receiver, for instance in the 80 m band between the countless telegraphy transmitters, fast gets an idea of what had been up in the air already 2000 years ago. No ancient city would build several temples on a single Acropolis, if only one single one could have been used.

All temples broadcasted with each time another carrier frequency because of different dimensions. For this reason the temples, which stood side by side, as a rule were dedicated different gods. An acknowledgement, "the air just being free", in addition hardly was possible, because of the often-found spatial distance between the temple installations and the respective oracle. Between the transmitter of the god Apollo in Didyma and the receiver, the oracle of Milet, for example lie approx. 20 kilometers.

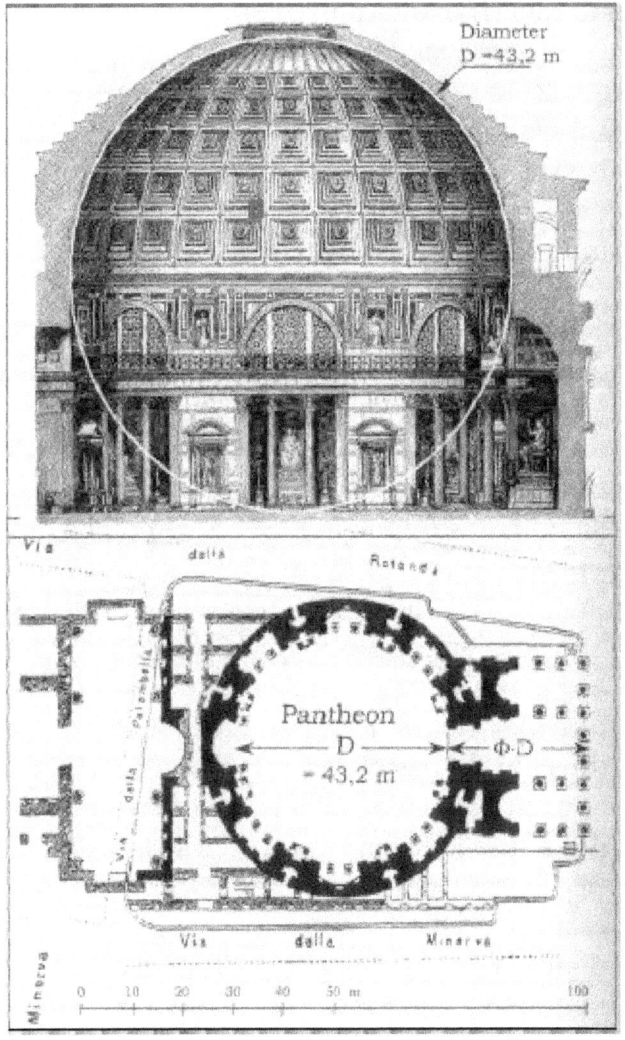

Fig. 30.9: The Pantheon in Rome, the „temple of all gods".
Diameter D = 43.2 m. Golden Proportion: Φ= 0.618
Pronaon-(atrium-)length: Φ·D= 0.618 · 43.2 = 26.7 m
Built under emperor Hadrian 118/ 119-125/128 A.D.

The only possible conclusion is that in antiquity there was broadcasted on all channels simultaneously regardless of other gods and their transmission frequencies. As is well-known there rather prevailed a situation of competition between the gods, since like today a large number of listeners meant great importance, influence and power and eventually also worship, more gifts and more receipts from broadcasting fees.

30.9 Broadband FM broadcasting technology

If I speak against a flat wall, then every point on the wall has another distance to my mouth. The sound waves thus aren't reflected simultaneously, what leads to big modulation distortions. Therefore the sound wall should be curved in such a way, that all signal paths are the same length (barrel vault, apse, etc.). In the case of point sound sources there results as an optimum a hemisphere, for instance a dome. The building hence even today tells us, which frequency and which modulation method had been put to use.

The architecture of sacral buildings, e.g. pointed arch or round arch, thus hardly has been a question of aesthetics. For his temple of Venus and Roma designed by himself emperor Hadrian had to listen to severe criticism among others of Apollodor of Damascus. The temple was too broadband for an AM transmitter, however with a modulation depth of just 11 percent not broadband enough for a phase modulated FM transmitter. Emperor Hadrian however also had the courage to build midst in Rome a temple calculated completely new and designed as a pure FM transmitter, the Pantheon, which means temple of all gods. In the language of the technician it is a transmitter for all frequencies.

This domed structure indeed doesn't leave out one single frequency. With a modulation depth of almost 100 percent it is designed for maximum loudness. With that the Pantheon uses all available frequencies, for which reason the name temple of all gods really is no exaggeration. Into the Pantheon exactly fits a sphere with a diameter of 43.2 meter. That corresponds to a minimum frequency of 3.47 MHz, situated in the range of the short waves.

The floor however is not domed, but horizontal. That, up to the basis of the dome, results in exactly half the height and a maximum frequency of 6.94 MHz. The construction ensures that between the simple and the double diameter any desired wavelength can be produced. Above the given maximum frequency of the basic oscillation the harmonic waves, which are produced as well and can't be avoided at all, are attached without a break. These occupy the wave bands up to the double, triple, quadruple frequency and so forth. For this and only for this reason a maximum frequency was chosen, which corresponds to exactly the double value of the minimum frequency. The operation takes place to the limit, where the transmitter would interfere with itself, in the way that the used basic oscillation would overlap its own harmonic waves. That then sounds like two people talking at the same time. The voices would be distorted out of recognition, as can't be expected else in the range of the harmonic waves.

The Pantheon has been planned and built as a phase modulated basic wave transmitter according to purely academic rules of Hadrian. The temple impressively demonstrates the precise engineering detailed knowledge of the Pontifex Maximus and his broadcasting priests in ancient Rome.

The official version of history is just a pack of lies; people have been spoon-fed this rubbish all their lives. Surely, COMMON SENSE can prevail over the blind following of a system that has been deceiving you. When I tell people about what has really happened in history they call me crazy, well I am not and I am an intelligent, observant person who will not be led on or deceived. QUESTION EVERYTHING.

Trust, the System of Deception

"All names of gods must be in "ALL CAPITAL LETTERS " All COVENS of gods must be in all CAPITAL LETTERS preceded by "of" to denote "possessed by god"; and must display their COVEN Sigil" Which became called the "CORPORATE LOGO" This is why "CHURCH and CORPORATION" must be in all CAPITAL LETTERS on legal documents to be legal.

Government Is Foreclosed from Parity with Real People – Supreme Court of the United States 1795 [Not the "United States Supreme Court"]

"Inasmuch as every government is an artificial person, an abstraction, and a creature of the mind only, a government can interface only with other artificial persons. The imaginary, having neither actuality nor substance, is foreclosed from creating and attaining parity with the tangible. The legal manifestation of this is that no government, as well as any law, agency, aspect, court, etc. can concern itself with anything other than corporate, artificial persons and the contracts between them." S.C.R. 1795, Penhallow v. Doane's Administraters (3 U.S. 54; 1 L.Ed. 57; 3 Dall. 54) - Supreme Court of the United States 1795

The Court is the synagogue. The Temple of Baal, enforcing Babylonian Talmudic Law. The gate (or bar) is the veil. {one enters to give sacrifice} The bench is the alter. The Black Robed Devil (the judge, administrative magistrate) is the high priest. {vicarius dei} The Attorney [from Latin, attorn = to twist or turn] is the mediator. {vicarius filii dei} The attorney's job is to move one into Roman 'Civil Law' Jurisdiction and then quickly into Code and Rule Pleadings (Babylonian Law); remember he is a devil, too.

The Law of the UCC: Uniform Commercial Code
Around 1930 the time of the great depression most governments of the world were bankrupt and as payments to Bankers the governments floated a bond against our future earnings by using our birth registrations as the collateral for our promise to pay Income tax is just their having educated you to pay the interest on the loan YOU lent THEM. When we access our Direct Treasury Accounts, those held at the BC/ FRB under our SINs/ SSNs, we will no longer have to work.

Strawman "LEGAL FICTION" Natural V's Artificial Person

There are two "persons" identified in law. These are "natural-person" and "artificial-person".
A natural-person is defined as "A human being that has the capacity for rights and duties". Note that the word capacity means the ability, but not the obligation for rights and duties.

An artificial-person is defined as "A legal entity, not a human being, recognised as a person in law to whom legal rights and duties may attach - e.g. a body corporate". Sometimes an artificial-person may be referred to as a CORPORATION, which is not always the same as an Incorporated Company. These subtle re-definitions are made in Statutes whenever the Government wants to change the meaning of the word.

There are many different types of artificial-persons, each with different duties. Here are a few different types of artificial-persons:
Taxpayer, Resident, Driver, Voter, Citizen, Home-owner, Officer.

Whenever you read any Law or Statute, you must be sure to check the meaning of the word "person" as it applies to that particular law. In order to implement slavery of its citizens and control them according to its whim, the Government had to invent a system that would not violate a human-being's fundamental rights, but would allow the Government to "own" everything produced or gained by its citizens.

The technique used by the Government was to create an artificial-person (referred to herein as a CORPORATION for emphasis) for every human-being in Australia. As creator of a CORPORATION, the Government can demand anything it wants from the CORPORATION. As a legal entity, a CORPORATION does not have feelings and cannot be hurt. It can be subject to slavery and complete domination by its creator and the CORPORATION must obey its creator.

So for every John Doe human-being in Australia, the Government created a JOHN DOE CORPORATION. Capital letters are used to represent CORPORATIONS and COMPANIES. Lower case letters are used to represent the name of the natural-person.

As a CORPORATION needs a business number, in order to do business, the Government assigns a unique business number to each JOHN DOE it creates. The creator (Master) can then track all activities of the Slave and claim ownership on all property and income of the Slave.

Finally the Government needs to appoint an Officer of the CORPORATION to run the day-to-day activities. Such a position requires a contract since the Officer will be held accountable for the actions of the CORPORATION. So, the Government tricks John Doe to become the Officer for the JOHN DOE CORPORATION by signing such contracts as Driver's Licence, Bank Accounts, Citizenship Cards, Passports, etc. In the Income Tax Act the Government just decrees that John Doe is the Legal Representative for the Officer of the JOHN DOE CORPORATION and the only contract involved is the annual Income Tax Return (yes it is a contract for one year) wherein John Doe gives his agreement as Officer of JOHN DOE for the previous year. Unfortunately John Doe does not know that he is an Officer for the JOHN DOE CORPORATION and must therefore follow the rules imposed upon JOHN DOE. Hence the confusion sets in because John Doe believes that he is JOHN DOE and therefore has to forfeit his rights and duties upon demand by the Government and its officials.

"Whereas it is essential, if man is not to be compelled to have recourse, as a last resort, to rebellion against tyranny and oppression, that human rights should be protected by the rule of law,"

THIS IS HOW THEIR SYSTEM OF FRAUD WORKS.
I know for a fact that from the creation of my birth certificate in New Zealand there was a corporate artificial identity created named WAYNE MURRAY THOMPSON. This artificial identity in ALL CAPITAL LETTERS goes through the government department concerned with Births Death and Marriages and in HER MAJESTY THE QUEEN IN RIGHT OF NEW ZEALAND Corporation / government then goes through the LAND TRANSFER ACT.

WAYNE MURRAY THOMPSON is LAND with a real monetary and wayne murray thompson is a natural person.

During this process WAYNE MURRAY THOMPSON is given a monetary value which the BANKS use when you illegally borrow money from them. I know for a fact that CREDIT CARDS and BANK LOANS are based on deception and fraud as it is my money they are giving me from the creation of my artificial identity. I have contacted the CEO of my bank asking to show the FULL ACCOUNTING for WAYNE MURRAY THOMPSON and they stand silent as they know they have committed fraud. THEY DO NOT ASK FOR THE MONEY BACK AS IT WAS RIGHTFULLY YOURS IN THE FIRST PLACE.
I know there was never any debt as all money concerning me can come from WAYNE MURRAY THOMPSON as that is what it was created for. YOUR LIFE IS PREPAID. THERE IS NO DEBT YOU ARE SLAVES.

President Lincoln's first inaugural address made on March 4, 1861. He stated:
"This country, with its institutions, belongs to the people who inhabit it. Whenever they shall grow weary of the existing government, they can exercise their constitutional right of amending it, or their revolutionary right to dismember it or overthrow it."

Go to Santos Bonacci "Your Soul Is Owned By the Vatican" on YouTube,
We need to wake up. The society we live in is nothing but a pyramid scheme of mind control.
Pope Boniface VIII in 1302 during the Inquisition, an "express trust" was created with a Papal Bull called "Unnam Sanctum".
It created this and it created the Roman Empire from Roman Cult Sorcery.
It has 3 different "Cestui Que Vie" trusts-
1. Romanus Pontifex
2. Aeterni Regis
3. Convocation

All law is Ecclesiastical Law. This law allows "Constructive Trusts" to be created.
These trusts demand you retire before 75, because that is outside the law of the trust to claim.
In a courtroom you have an Administrator, Trustee, Executor and a Beneficiary.
A person taken in to court is the Beneficiary. Your Birth Certificate is nothing more than the title of the Trust, the Judge in a court, who is basically an Administrator, asks for the person's name, if you give your name, the one on your birth certificate, then you are allowing the Judge (Administrator), to swap roles with you as the beneficiary.

A constructive trust is defined as – "Constructive trusts in English law are a form of trust created by the courts primarily where the defendant has dealt with property in an "unconscionable manner", but also in other circumstances; the property will be held in "constructive trust" for the harmed party, obliging the defendant to look after it."

Once they have the name, you have given the fiction of the Cestui Que Vie Trust, life. It is nothing but modern slavery; it allows the Stock shares, which are generated by your birth certificate, to be taken from you.

The Government has to supply evidence to a Monetary Fund that proves there is a demand for money; your Birth Certificate is proof of that. If you have "dealt with property in an unconscionable manner", then the Judge, or the state, has the power if you identify with the name on the trust, to take this property off you. The Judge receives a commission, the public purse is relieved of an expense, but you still exist as demand, so whether or not you are able to contribute, if you are part of the trust, you are not permitted to be a beneficiary.

The Clerk is the Trustee, the Prosecutor is the Executor. The Prosecutor has the liability, they create the summons to a hearing. Once you admit to being the beneficiary to the trust, then the Executor becomes the beneficiary and you become the Executor and you execute your own sentence. If you don't identify with the trust, then the Executor has to pay, as they have bought the proceedings to claim.

The Ecclesiastical factor means that a courtroom is about the "Sacrament of Penance." This means that you are a "Sinner", you have been a source of funding for the Stock Exchange and the system has now capitalized on your felony, your predisposition as a sinner, so you are no longer permitted to handle the property of the state in an unconscionable manner, the Ecclesiastical Trust in the court represents Church Law, the Judge, Clerk and Prosecutor and even your own Lawyer or Solicitor, are representatives of the Church, they are in business with God, Priests. If you sin, they are saying that they can offer forgiveness by getting you to confess and identify with the trust and that you will be "forgiven" by God, if you give them the money that your existence and name permits the banks and the Governments to make money "out of thin air".

It is an Administration of credits from the Kingdom of Heaven. This "Administration" is "Penance".
It only works with confession; it runs like the Church and is a glorified confession booth.

You basically through your identifying with the trust, the accuser, the person accused and the witness and the Priests give the Judgement and the sentence and has the power to forgive sins.

They manage the indulgencies, which are sins committed after forgiveness.

They have monetized sinning, which has got to be a problem in our disastrous economy. Instead of using your money in this corrupt trust to solve problems with technology and make resources of the Earth every human's inheritance, they instead create legal fictions like this.

A warrant is only in place to indemnify the Queen from being sued if they lose, a warrant also allows a "Writ", which is basically a "Rite", which is a religious term for a spell, to be sealed and indemnify the Monarchy. The basic truth is that it is illegal to send people to jail; it is only this inhumane legislation of reinterpreted madness that permits fraud.

A Writ is also an indulgence, but because of all this reinterpreting, no one picks up on the madness of it all, the piracy.

Prosecutor can be broken down in Latin to mean – "Representing one's own flesh, or a person who is claiming to be you, making a false accusation".
If the Prosecutor can't get you to swap roles, then they have to pay the liability and they receive no commission from your trust account. You are nothing more than a bank account.

The way out of this is just as insane as the trust.
Instead of having a birth certificate, you can have an Ecclesiastical Birth Certificate called a Live Born Record. This is an independent system that doesn't identify with the trust. It means that you have documented proof you are divine and so you can't sin!

By this fact, if you have a birth certificate, then your soul is owned by the state and ultimately, the Vatican!
You can also say this in court, you don't have to go to prison, no one does, it is all illegal.

When the name of the trust is called for instance the "John Doe" trust, you can say, "Are you saying that the trust we are now administering is called the John Doe trust, your Honour?"

The very mention of this knowledge will put the fear of God into the Judge.

Don't ever identify with the trust.

"We can now establish that the trust is the name of a trust 'Not a live man', what is your next question your Honour?" Judge – "What is your name?"

You must be very careful not to identify with the name of the trust because doing so makes us the trustee. What does this tell you about the Judge? If we know that the Judge is the trustee, then we know that the Judge is the name, but only for this particular constructive trust.

As you will notice the Judge will become frustrated with the refusal to admit being the name, that they will issue a warrant and as soon as the man leaves they arrest him, how idiotic is that?

They must feel foolish for admitting that John Doe is not in the Court, so I'm issuing a warrant for his arrest and as soon as the man they just admitted is not there to be arrested because he is there.

They must get us to admit to being the name or they pay and we must not accept their coercion or we pay because the Judge is the Trustee, a precarious position.

The best thing to say in that case is "John Doe, is indeed in the court, your Honour." Point to the Judge.

"It is you, as trustee, you are John Doe, today, aren't you?" Why not? We are men and women, we are not persons. We have Dominium.

During their frustration over not admitting to being a trust name, the trustee and/or the executor of the trust, we ought to ask who they are.

"Before we go any further, I need who you are." Address the Clerk of the court.

"The trustee for the Cestui Que Vie trust owned by (STATE/COUNTY/PROVINCE), are you the Cestui Que Vie trustee who has appointed this Judge, as a trustee for the Cestui Que Vie trust owned by (STATE/COUNTY/PROVINCE)?"

"Are you the Cestui Que Vie trustee who has appointed this Judge as Administrator and trustee of the constructive trust case (case number(s))?"

"Did you also appoint the prosecutor as executor of this constructive trust?" Then point to the Judge.

"So you are the trustee,", then point to prosecutor, "you are the executor, are you not? And I am the beneficiary."

"So now we know who's who, I as the beneficiary, I authorise you, to handle the accounting, and dissolve this constructive trust."

This is the power we have, dissolve this confession; I'm not into sin, I'm Divine.

You are a spirit, they view you as a dead soul, lost at sea, minor, and incompetent and award of the state.

"I now claim my body, so I am collapsing the Cestui Que Vie trust, which you have charged, as there is no value in it. You have committed fraud against all laws likely we will not get to hear that before the Judge will order. Case Dismissed!"

Or even more likely – the prosecutor will say clutching his cheque book "We've withdrawn the charges."

There are 100,000s of people doing this.

When you go to court, magistrates, you are under UCC law, this means you can only be fined.

When the Judge has a recess he can change it to Canon Law or Maritime Admiralty Law, this can sentence you to prison. Admiralty Law is in play as you are then seen as lost at sea, you are officially stock on a port and they now have the power to store you in a warehouse (prison). If the Judge orders the Bailiffs to throw you out of court, then you can say "Don't do that, you are dishonouring a court official.", because the documents have been handed to the bailiff.

If the Judge goes to leave the court for recess, then you must acknowledge – "The Judge has jumped ship, for the record, he has abandoned ship and I as Sovereignty in this court take control! Case closed! With Prejudice!"

If the Judge says they are going to have a recess, you can decline it, as it is an offer. "Your Honour, I don't consent." If it is adjourned, then it stays in the same court, so just follow the above.

Always relay that "I seek leave for an Interlocketary appeal, on a matter of law."

If you say this, the Judge will be reluctant to go to appeal because he will be losing out on commission.

If you don't give the name and say your name is "John Doe", which is the title that is given to a dead person, then the Judge has to create the trust in that name, but because John Doe is not a live man, he can't be in the court, so they have to issue a warrant to arrest him, but they can't because he doesn't exist.

Using the information above means that you are not consenting for the Judge to swap roles with you, if you don't confess, then you can't sin, it is all an elaborate fiction. This confusing system is basically founded on the Church that Jesus and God are the only ones who can forgive, they represent this fiction and it is saying "You are a filthy rotten sinner, you can't be saved on your own God, and you go to Prison and give me some money".

The birth certificates are the names of the trusts, giving that name gives the fictional trust life.

The truth about the Bible is that it says that the Kingdom of God is within you, not some mythical omnipotent, hierarchical, patriarchal entity; they have capitalized on this to steal from the poor. By not giving them the tools to create the trust allows you to be Divine, you are a spirit, not a person, if you are divine, you are not a sinner.

Watch the Santos Bonacci Lecture and you will see the trust and how it is in everything.

A spirit can't have a name, the church can't own anything, this is why this madness is in operation, if we are awake, we don't have names, we have a soul, keep people asleep and hide all this fiction and call it law, then you will have a name that they can buy and sell like stock to put in a warehouse (prison), they can literally own your soul, as you are saying that you ARE a sinner and you are consenting to confession and penance (hearing and sentencing).

You stand in a dock in court, you are on the port, a lost soul, lost at sea, incompetent etc.....this is how they view you, so you can say you are John Doe, not a live man, the trusts can only charge a trust account with a living person. When you don't allow yourself to give them your soul, you are proof that you come from heaven, heaven is in you, and so you can then legally claim your body back!

It's insane law made by insane people. They know they can't just take things, so they have found a way to get you to give it to them.

The property that is abused in an unconscionable manner is the money in the trust, if you confess, you admit that you cannot look after this property and effectively prosecute yourself and hand over everything that is yours to the state, you are the "award of the state".

You are an incompetent - so by law, if you stand in court, you're an idiot, this is how these box tickers work, and it is sick.

If you don't give a name, they can't take anything from you. Remember a warrant is only a technicality to protect the state from prosecution if they are wrong, it is illegal, but they have made this stuff up so you give it to them, without you even realizing it.

This is how the state makes money, they make things so expensive and raise taxes so that you are much more likely to commit crime, when you do, they send you to prison, but they can still claim from monetary fund's under the name on the Cestui Que Vie Trust.

If you have to go to court, say what is above and you will win, as they will have to lose commission and pay the liability for the charges bought before the court. You can't be prosecuted if you don't give the name on the trust, if you don't do that, then the only person who can be prosecuted is the Judge, and the Prosecutor will have to pay the charges as they are liable for the whole thing. You can take this with you, you don't need to remember it, it is worth knowing, if you show even an inkling of how a constructive trust legalizes fraud, they will try to dismiss the case before the charges have to be made. Either you pay and they get commission, or they pay and lose commission.

It's all sales of souls; literally, it's a curse on Humanity. With an ecclesiastical deed poll, you will have a new trust number that doesn't have Roman numerals and you have documented proof that you are a spirit.

Remember you are a dead soul, lost at sea, let them know you know this and claim your body back! It is insane, but completely true, the Vatican is the owner of all land; they have taken the wealth from us all, when we register our birth certificate, and nurses get a commission.

Expose this false institution, the revelation says that they operate just like this Church; the baptism is their way of pulling wool over our eyes, be nature people, not sheeple. Also when a Judge walks into a court room they say, "All rise." If you stand this is seen as you entering into a CONTRACT and accepting their Authority over you, and you don't want that. When a Police officer reads you your rights the last thing they say is "Do you Understand" Don't be fooled into thinking it means do you comprehend, it means "Do you stand under my Authority." Just say "no" then you do not Contract to them. The same is with your "DRIVERS LICENCE" when they ask to see your licence and you voluntary show it to them this contracts you to the right to get a fine. We are all travelers on the road unless you earn money by DRIVING; we all have the right to travel.

Kansas" – it was now "KS," artificial corporate venue of the bankrupt United States, i.e., the "Federal Zone," newly established "federal property," and Dorothy and Toto were "in this state" (see definition for "this" state). In the 1930's the all-capital letters corporation sole, straw man, newly created artificial aspect of the former American sovereigns, had no brain – and Americans were too confused and distracted by all the commotion to figure out that they even had a straw man. The Scarecrow identified his straw man persona for Dorothy:

"Some people without brains do an awful lot of talking. Of course, I'm not bright about doing things."
THE WIZARD OF OZ
A motion picture made in 1939. Metro-Goldwyn-Mayer. Book by Frank Baum

Just as you can read between the gory lines in the newspaper on any day and discover clues issued by the Powers That Be – if you look hard enough – as to what is actually going on, such notice can also be found in lighter fare, like the movies. Such a movie was THE WIZARD OF OZ, an allegory for the new state of affairs in America in the 1930's following the stock market crash and factual bankruptcy of the US government immediately thereafter.

The setting was Kansas; Heartland America, and geographical centre of the USA.

In comes the twister, the tornado, i.e., whirling confusion – the stock market crash, theft of America's gold, US bankruptcy, the Great Depression – and whisks Dorothy and Toto up into a new, artificial dimension somewhere above the solid ground of Kansas. When they finally land in Oz, Dorothy comments to her little companion: "Toto, I have a feeling were not in Kansas anymore."

That's right. After the bankruptcy, Kansas was no longer just "plain old Ka

And in his classic song, "If I Only had a Brain," the Scarecrow/Straw Man succinctly augured:
"I'd unravel every riddle, for every 'individdle,' (individual), in trouble or in pain."

Translation: Once one discovers that his straw man exists, all political and legal mysteries, complexities, and confusions are resolved – and once one takes title to his straw man, he will no longer be in trouble legally or be damaged legally.
The Tin Man was a hollow man of metal – a "vessel" or "vehicle", newly created commercial code words for the corporation sole straw man. One of the definitions of "tin" in Webster's is "worthless; counterfeit." Just like the Straw Man had no brain, this Tin Man vessel had no heart. Both were "artificial persons". The Tin Man also represented the mechanical and heartless aspect of commerce and commercial law. Just like they say in the Mafia: "Nothing personal, it's just business." The heartless Tin Man also carried an axe, traditional symbol for God – i.e., modern commercial law – in most earlier dominant civilizations, including fascist states.

The "Ace", etymologically related to "axe" in a deck of cards, representing God, is the only one above the King. One of the "Axis" Powers of World War II, Italy, was a fascist state. The symbol for fascism is the "fasces," a bundle of wooden rods containing an ax with the blade projecting. The fasces may be found on the reverse of the American Mercury-head Dime (the Roman deity Mercury was the God of Commerce), and on the wall behind, and on each side of, the Speaker's podium in the US Senate. At the base of the Seal of the US Senate are two crossed fasces.

The Lion, "king of beasts," i.e., king of the goyim denigration in itself – representing the once-fearless American People, had lost his courage. After your first round with the UCC-constituted IRS "defending" your straw man/vessel/vehicle/all-capital letters name/artificial person, you probably lost some of your courage too. You didn't know it, but the IRS deals with you strictly under the laws of commerce. Just like the Tin Man, commerce is heartless.

To find the Wizard you had to "follow the yellow brick road," i.e., follow the trail of America's stolen gold and you will find the thief who stole it. In the beginning of the movie the Wizard was represented by the traveling professor/mystic whom Dorothy encountered when she ran away with Toto. His macabre shingle touted knowledge of "The Crowned Heads of Europe, Past, Present, and Future." Before the bankers stole America, they had long since dethroned the Christian monarchies of Europe and looted their kingdoms. With a human skull peering down from its painted perch on his wagon behind him, the good professor lectured Dorothy of Osiris and Egypt.

When Dorothy Gale and crew emerged from the forest and arrived in Emerald City, the city of green (the new "fiat money", or money by decree, Federal Reserve Notes), they were serenaded by the Munchkins on the glory of the Wizard's creation:

"You're out of the woods, you're out of the dark, you're out of the night, Step into the sun, step into the light, the most glorious place on the Earth!"

The foregoing jingle abounds with Illuminist-Luciferian symbols and metaphors, e.g., darkness and light.

The Wicked Witch of the West made her home in a round, medieval watchtower, ancient symbol of the Knights Templar of Freemasonry, who are given to practice witchcraft and who are also credited as the originators of modern banking, circa 1099 A.D. The Wicked Witch of the West was also dressed in black, the colour symbolizing the planet Saturn, sacred icon of the Knights Templar, and the colour of choice of judges and priests for their robes. Who was the Wicked Witch of the West? Remember, in the first part of the film her counterpart was "Miss Gulch", who alleged that Dorothy's dog, Toto, had bitten her. She came to the farm with an "Order from the Sheriff" demanding that they surrender Toto to her custody. Aunt Em answered Miss Gulch's allegations that Toto had bitten her:"He's really gentle. With gentle people, that is."

Could "gentle" be a metaphor for "Gentile"? When Dorothy refused to surrender Toto, Miss Gulch threatened: "If you don't give me that dog I'll bring a damned suit that'll take your whole farm!"

Today, 70% of all attorneys in the world reside in the West – America, to be exact – and 95% of all law suits in the world are filed under US jurisdiction. The Wicked Witch of the West and Miss Gulch, my friends, represent judges and attorney; i.e., the American legal system (including the attorney-run US Congress), executioner and primary henchman for transferring all wealth in America – everything – from the people to the banks and the government. Miss Gulch wanted to take Toto. What does the word "toto" mean in "attorney language," i.e. Latin? "EVERYTHING"

Dorothy and the gang fell for the Wizard's illusion in the beginning, but soon wised up and discovered the Wizard for what he was: a confidence man. When asked about helping the Scarecrow/Straw man, among some other babblings about "getting a brain" and "universities", etc., the Wizard also cited "the land of E Pluribus Unum," which is Latin for "One out of many," i.e., converting the many into one = New World Order, or Novus Ordo Seclorum. He also proudly revealed/confessed that was:

"Born and bred in the heart of the Western Wilderness, an old Kansas man myself!"

The bankers did pretty well in Europe, but as the Wizard pointed out, they made a killing in the "Western wilderness" – America – with the theft of American gold, labour, and property from the – quoting John D. Rockefeller – "grateful and responsive rural folk" who populated the country at that time.

When Dorothy asked Glenda, the Good Witch of the North, (Santa Claus, and Christianity) for help in getting back to Kansas, Glenda replied,

"You don't need to be helped. You've always had the power to go back to Kansas."

Translation: you've always had the right and power to reclaim your sovereignty, you just forgot. The actual act of reclaiming your sovereignty – remedy – a simple UCC-1 form to the Secretary of State can be completed from scratch in a few hours' time.

America and Americans have intimate first-hand knowledge of the heartless mechanics of the laws of commerce, religiously applied by the unregistered foreign agents of the IRS. The IRS, collection agency for the private Federal Reserve Bank, was constituted under the UCC in 1954 and has been operating strictly in that realm ever since. (The UCC was formerly adopted as law in 1963) Everything worked out for Dorothy, i.e., the American people, in the end and she "made it home." Meaning: there is remedy in law. It is there – it was just encoded and disguised and camouflaged. Fortunately, the code has been cracked, and we're home free, just like Dorothy. Why continue as a slave, to be further conned by the confidence men? Like Dorothy said, "There's no place like home" – and there isn't! There is nothing like sovereignty for a sovereign! We have commercial remedy in the Redemption process. Will you continue to sit back and take in the Wizard's light show, or will you wise up like Dorothy and "look behind the scenes?"

Note: in the movie, Dorothy's shoes were ruby slippers, in the book written by Frank Baum, her slippers were silver. It has been said that the movie slippers were ruby, because if they would have been silver, the scheme may have been figured out.

From parts of Chapters 4 and 12 OF "TOP SECRET BANKER'S MANUAL"
What Bankers Fear

Tom taught over 2,000 CPAs nationally on appraising businesses and testifying in court as an expert witness. Tom owned and operated his own CPA firm and business brokerage business for about ten years. After one of the seminars in Pennsylvania at a Holiday Inn, Tom talked to a controller (top accountant) for a major bank. In a private conversation, Tom thought he would see if he could get a reaction out of this accountant. Tom said to the controller, "You know that all your bank loans are a fraud." Without hesitation the controller agreed. Tom said, "Aren't you afraid that you will go to jail." The controller responded, no.
He then explained how banks create money and he who owns the money controls the judges, lawmakers and the media. He explained how advertising money, loans and direct bank ownership and how banker's political contributions control the politicians and the laws and how money controls the media. If a politician votes against the bank, the bank heavily funds their opponent next election so that the bank politician wins. All the politicians know that they need the bank's media and money to get elected. He even boasted how the bank controls the F B I. (Get the idea of why they took away rights if they call someone a terrorist?).

He then said, " I f someone put together a brochure and passed it out in mass, I would immediately, permanently leave this country. If the American people ever figure out what we have done to them, they would put all of us bankers, judges, sheriffs, and lawmakers in jail. "He then laughed and said, "The American people are too stupid to figure out what we have done to them, they will never be able to explain this in court." He let Tom know how foreclosures are very profitable and when the bank helps the judges, politicians, and sheriffs get the profitable foreclosures. The government agents in the bankers' pocket have very profitable investments.

The bankers and politicians call it good business. They represent their personal investments, not the people that elected them. Currency trading is also very profitable. Some government agents helping the bankers get 100 percent profit a month on their investments. He explained how the government agents sold their souls to the bankers all for the love of money.

The bankers' own secret manual that is truly for the bankers, shows that the bankers hate it when people claim "fraud in the factum" (fraud in the execution). Remember the law in U S C Title 5 Administrative Procedures Act? The nation is bankrupt so we are under administrative law and that is the law of "notices". Remember how the IRS and the banks always give you a notice? You need to do the same. Notice them asking what the terms of the agreement are — the agreement that they wrote.

When they refuse to tell you, the theory is that you can claim "fraud in the factum".
Chapter 12- Ultimate fear of Bankers

The banker can only say that there is an agreement and that you owe money. The banker cannot show you the original promissory note after it was altered. The banker fears that the borrower might claim that the agreement says that the borrower can repay using another IOU - promissory note payable in the same specie of money, money equivalent or credit or funds or capital that the bank or financial institution used per GAAP to fund the loan, thus ending all interest and liens. This would allow the borrower to discharge the loan, and all interest and liens.

The banker knows that if this is claimed, then you could repay not with cash or a check, but with a promissory note, also payable in the same specie of money the bank used to fund the loan, per GAAP, thus ending all interest and liens. If the banker insists that you pay the note, you ask the banker to sign the back of the note, and you replace it with another note.

The banker fears that you viii claim that the original contract was altered and stolen and that there was at addition to the agreement with the following items l)The intent of the agreement is that the original party who funded the alleged loan per the bookkeeping entries is to be repaid the money. 2) The bank or financial institution involved in the alleged loan will follow GAAP. 3) The lender or financial institution involved in the alleged loan will purchase the promissory note from the borrower, 4) the borrower does not provide any money, money equivalent, credit, funds or capital or thing of value that a bank or financial institution will use to give value to a check or similar instrument.

5) The borrower is to repay the loan in the same specie of money or credit that the bank or financial institution used to fund the loan per GAAP, thus ending all interest and liens, and 6) the written agreement gives full disclosure of all material facts.

Do you see the bankers fear? If the banker claims item number 1 is false, then it is a swindle. If item number 2 is false, then it is illegal. If item number 3 and 4 is false, the bank invested nothing. It was stolen or paid nothing for it and you funded the loan. If number 5 is false, then the bank admit it is only a moneychanger and charged as if there was a loan. if number 6 is false then they agree that they concealed material facts. How can the bank claim that the items are not part of the agreement? The banker knows that if this is claimed, the banker must show the original note. If the banker claims that he only has a copy, the borrower could claim that the additional part of the agreement is missing with Items I to 6. Now one is only arguing the agreement - not the banking system.

The banker must discuss GAAP and bookkeeping entries and items 1 to 6 are the last thing that the banker wants talk about, Imagine the banker's fear if the borrower sent a promissory note to repay the loan, claiming that the agreement allows it. Imagine sending in a check to repay the mortgage to be applied to the la note you sent. Imagine the potential lawsuit for the banker breaching the agreement and the banker cannot claim that items 1 to 6 are not part of the agreement.

The borrower says. 'How can I claim this?" The bank is incorporated and claims that they follow the law - GAAP- with full disclosure in their agreement and without false and misleading advertising. They claim that they lend you their money - how can they claim differently? Bankers fear that they will have to explain the agreement GAAP and who funded the loan. The banker wants you to argue the banking system, which means you will lose in Court. They do not want you to Claim breach of agreement and claim items 1 to 6 are part of the agreement and they would have to claim items I to 6 are not part of the agreement. Bankers understand that if they refuse to show the original agreement.

The borrower can claim that the copy is forged because it leaves out items I to 6. Bankers fear that borrowers may say "fraud In the factum", claiming that the Items 1 to 6 are concealed or there Is a forged document leaving the items Out. Who cares who funded the loan You care because it changes the cost and risk of the loan, if there Is nothing wrong with stealing and counterfeiting, then why do we send those kind of people to jail?

Alter you send all the notices, ask for a closing statement to discharge the debt. Then offer to discharge the debt with cash or same specie of money as discussed earlier, providing that the bank returns the original unaltered note at time of payment. They will refuse. This allows you to sue. This has led to many wins.

IN THE LAW, WHAT IS A "HUMAN BEING"?

Clinical Laboratories v. Connecticut Blue Cross, Inc., 31 Conn.Sup. 10, 324 A.2d 288, 291. See Per se violations.

Persequi /pərsəkwāy/. Lat. In the civil law, to follow after; to pursue or claim in form of law. An action is called a "jus persequendi."

Per se violations. A term that implies that certain types of business agreements, such as price-fixing, are considered inherently anti-competitive and injurious to the public without any need to determine if the agreement has actually injured market competition. See Per se doctrine.

Person. In general usage, a human being (i.e. natural person), though by statute term may include a firm, labor organizations, partnerships, associations, corporations, legal representatives, trustees, trustees in bankruptcy, or receivers. National Labor Relations Act, § 2(1).

Bankruptcy Act. "Person" includes individual, partnership, and corporation, but not governmental unit. Sec. 101(30).

Resident alie the meaning clauses of t Enterprises, D.C.N.Y., 41 Unborn chil teenth Amen v. Wade, 41(147. A fetus protection u row v. Cliffo A viable us alive but for son" within Simmons v. F.Supp. 529. of remedies s sue after his 493 P.3d 130

Persona /persō virtue of wh certain dutie may unite m

person An indispensable word with varied, overlapping meanings. Often used without definition, as in the U.S. Constitution (Arts. I, II, III, IV; Amends. IV, V, XII, XIV, XXII). Defined, and redefined, in an endless succession of special purpose statutes, with no assurance to the profession that this is the *person* you thought you were talking about. The definitions here give an overview of current usage. This omits a whole list of historical horrors in the ugly shadows of slavery, racism, and sexism.

1. <u>a human being</u>—without regard to sex, legitimacy, or competence. This *person* is the central figure in law, as elsewhere, characterized by personal attributes of mind, intention, feelings, weaknesses, <u>morality common to human beings</u>; with rights and duties under the law. This is the *person*, <u>sometimes called an</u> *individual*, and often referred to in the law as a <u>natural person</u>, as distinguished from an *artificial person (sense 3).*

McIlinioff's Dictionary of American Legal Usage, 1992

Black's Law Dictionary, Fifth Edition

NATURAL PERSONS
See, also, Abuse of a Natural Person.

For purpose of statute protecting certain property from postjudgment remedies, and therefore from prejudgment attachment, "natural person" means human being, and not artificial or juristic person. Shawmut Bank, N.A. v. Valley Farms, 610 A.2d 652, 654, 222 Conn. 361.

Employee-user of goods purchased by employer is a "natural person" within the meaning of the

McIlinioff's Dictionary of American Legal Usage, 1992

NO WORD
"Human being"

Black's Law Dictionary,
Fifth Edition

LAW DICTIONARY
with
PRONUNCIATIONS

by
JAMES A. BALLENTINE
Professor of Law in the University of California

To find a definition of a "human being", you have to go back to 1948.

1948 EDITION

THE LAWYERS CO-OPERATIVE PUBLISHING COMPANY
ROCHESTER, N. Y.

huisher (wee'shay). Same as huissier.
huissier (wee'she-ay). (French.) A court usher; a process server.
hullus (hul'lus). A hill.
humagium (hu-mā'ji-um). A humid or moist place.
human being. See monster.
human body. See body.

Law Dictionary with Pronunciations, James A. Ballentine, 1948 Edition

monster (mon'stėr). A human-being by birth, but in some part resembling a lower animal. "A monster . . . hath no inheritable blood, and cannot be heir to any land, albeit it be brought forth in marriage; but, although it hath deformity in any part of its body, yet if it hath human shape, it may be heir." 2 Bl. Comm. 246

monstrans de droit (mon'stranz duh drwo) A showing or setting forth of the right

Law Dictionary with Pronunciations, James A. Ballentine, 1948 Edition

Code is Set in Stone - Before the bankruptcy of the established nation states, [invocation of a debt-based system of finance], men and women sovereigns were personally accountable for their actions in courts that were set up to accommodate disputes among sovereigns (such as the original common law in England before 1066). Now, the people's straw men, who are legally owned property of the system, are the "accountable parties" as adjudicated and enforced by the system. Such formerly sovereign men and women are now personally accountable because they are inextricably joined with their straw men which are owned via implied contract by a handful of arch-charlatans. Courts today are set up to deal only with straw men. We, their unfortunate counterparts, are merely "along for the ride." A sovereign (real being) has no place in a contemporary court (commercial, dealing in artificial persons), and cannot be legally accommodated. Only if the current debt-based system of finance and government is rectified will we ever get back to a sane and just basis for resolution of disputes and a sound civilization.

Your Moral Code is Primary - As a great Indian sage stated the matter: "Seek the highest first." This means, inter alia, to maintain your integrity and ethical behaviour. In practice this requires that you keep your word, honour your contracts, and not depart from your principles. A fundamental flaw in Man's thinking is the notion that he can cheat moral or natural law (usually by trying to cheat others) and get away with it. Every such attempt generates inexorable cause/effect consequences, all man's philosophies, systems, and cleverness to the contrary notwithstanding.

The entirety of our predicament is due to failure to live in harmony and accord with moral and natural law. The current system is the cause/effect result of our own folly. All governments are expressions of, and exist by virtue of, the people's irresponsibility, ignorance, laziness, larceny, and surrender of personal power, freedom, and autonomy in exchange for "being taken care of." In other words, every government exists due to the express will, as well as implied default of, the people (combined with the willingness of the ruthless to accept and manage the surrender of the people's power to the fictitious, artificially created, "government"). As Joseph de Maistre noted: "Every country gets the government it deserves."

What Can People Do? -To start with you must declare that you and your strawman are not one and the same. This is done via a very carefully worded affidavit, a "Statutory Declaration of True Name". These affidavit states exactly what your correct name is and that you are not to be confused with or represent in any way any of the straw man versions of your name. Next you will actually create an entity that takes on the names of your strawman. That is to say you will literally bring your straw man to life. Your strawman will be an entity officially created under New Zealand law having its own legal status. The benefit is that you will not have to try and convince anyone that you are not one and the same as your strawman. It will be self-evident. The strawman will have its own director and owner. The director and owner will represent the entity in all matters. The place for service and official address will be the same as the man or woman the entity is created to protect.

For example: Mr. John Frank Doe and Mrs. Jane Mary Doe of 22 New Street, Newmarket, Auckland, wish to protect themselves. This is what they do: Mrs Jane Mary Doe forms a company called JOHN FRANK DOE LIMITED to protect her husband. She then registers several names as "doing business as" these are the entities' registered trademark - names, registered at the Companies Office. She registers: JOHN F. DOE, JOHN FRANK DOE, J. F. DOE, JOHN DOE, John Frank DOE, John F. Doe, J. F. Doe, John Doe, DOE John Frank. Mrs Doe registers 22 New Street, Newmarket, Auckland as the place for service etc.. The newly formed entity will conduct no business at all! It will have no IRD [Inland Revenue Department] number and no bank account. It will essentially be a shelf company that exists in name only to prove legally that the entity exists and that John Frank Doe is not that entity. Mr Doe will do the same for Mrs Doe.

Pre-Paid

Pre-paid is very simple. The entire economy is pre-paid. Look at it this way: We have a car sitting on a dealer's lot. You walk up to buy the car. Does the dealer ever tell you "I am glad you are going to buy this car because we have to find out how we are going to pay for this car to be built." No is the answer you would get, but that is exactly what they are doing when you go to the bank to get a loan. When do they ever build something and then talk about how they are going to finance it to be built. The product was paid for when the contract was put in place to collect the industrial recourses through the Army Corp of Engineers, EPA, DOT, and OSHA in Flint, Michigan to build it. Even more precisely, the item was paid for when the census did a per-capita poll to identify how much money those agencies should put into the economy based on our productivity, (unfortunately take a quick look at Marxism and Keynesian Economics to make a connection with your worth and your previous status). Now everybody with a head (per capita) raise your hand. Good they loaned against you to finance the operation, that is the "Principal Account." Making the item pre-paid for the acceptor. This is another reason why you are the principal. The principal reason you are Pre-Paid is because Christ's acceptance of the sins in the Garden of Gethsemane and His death on the cross, created the Pre-Payment of all your liabilities both temporal and spiritual because they are inseparable because I wasn't here two thousand years ago but My sins were pre-paid on the condition that I accept the Redeemer. You are the source of economic production being the principal and your interest accruing from you i.e. a per-capita census statistics was pledged as the collateral to be the sponsor of the monetary systems' credit. That is why when interest that accrues from the principal gets returned (tax returned) to the principal, there is a decrease in tax liability (a deduction). The vendor is paying his taxes to you. That is why it is a tax matter. Tax is just a return of the interest to the principal.

Conclusion

**This is the face of anonymous.
Guy Fawkes**

Guy Fawkes a Catholic dissident along with 12 co-conspirators spent months planning to blow up King James I of England during the opening of Parliament on November 5, 1605. But their assassination attempt was foiled the night before when Fawkes was discovered lurking in a cellar below the House of Lords next to 36 barrels of gunpowder.

We need not hide behind a mask, we should stand tall and stand our ground and don't believe a word our governments are saying. They have all been corrupted. The person you vote for is not the one in control.

JFK speech on New World Order.
Selected Transcript of John F. Kennedy's Address before the American Newspaper Publishers Association, April 27, 1961: November 22, 1963 John F. Kennedy was assassinated

"The very word "secrecy" is repugnant in a free and open society; and we are as a people inherently and historically opposed to secret societies, to secret oaths and to secret proceedings. We decided long ago that the dangers of excessive and unwarranted concealment of pertinent facts far outweighed the dangers which are cited to justify it.
Even today, there is little value in opposing the threat of a closed society by imitating its arbitrary restrictions. Even today, there is little value in insuring the survival of our nation if our traditions do not survive with it. And there is very grave danger that an announced need for increased security will be seized upon by those anxious to expand its meaning to the very limits of official censorship and concealment. That I do not intend to permit to the extent that it is in my control.

And no official of my Administration, whether his rank is high or low, civilian or military, should interpret my words here tonight as an excuse to censor the news, to stifle dissent, to cover up our mistakes or to withhold from the press and the public the facts they deserve to know.

For we are opposed around the world by a monolithic and ruthless conspiracy that relies on covert means for expanding its sphere of influence--on infiltration instead of invasion, on subversion instead of elections, on intimidation instead of free choice, on guerrillas by night instead of armies by day.
It is a system which has conscripted vast human and material resources into the building of a tightly knit, highly efficient machine that combines military, diplomatic, intelligence, economic, scientific and political operations.

Its preparations are concealed, not published. Its mistakes are buried not headlined. Its dissenters are silenced, not praised. No expenditure is questioned, no rumour is printed, and no secret is revealed.

Without debate, without criticism, no Administration and no country can succeed-- and no republic can survive. That is why the Athenian lawmaker Solon decreed it a crime for any citizen to shrink from controversy.

And that is why our press was protected by the First (emphasized) Amendment-- the only business in America specifically protected by the Constitution-- not primarily to amuse and entertain, not to emphasize the trivial and sentimental, not to simply "give the public what it wants"--but to inform, to arouse, to reflect, to state our dangers and our opportunities, to indicate our crises and our choices, to lead, mould educate and sometimes even anger public opinion. This means greater coverage and analysis of international news-- for it is no longer far away and foreign but close at hand and local. It means greater attention to improved understanding of the news as well as improved transmission. And it means, finally, that government at all levels must meet its obligation to provide you with the fullest possible information outside the narrowest limits of national security.

And so it is to the printing press--to the recorder of man's deeds, the keeper of his conscience, the courier of his news-- that we look for strength and assistance, confident that with your help man will be what he was born to be: free and independent."

Read More: http://www.knowledgeoftoday.org/2011/12/new-world-order-exposed-john-f-kennedy.html

JFK *vs.* The Federal Reserve

http://ecclesia.org/

On June 4, 1963. a virtually unknown Presidential decree. Executive Order 11110, was signed with the authority to basically strip the Federal Reserve Bank of its power to loan money to the United States Federal Government at interest. With the stroke of a pen, President Kennedy declared that the *privately owned* Federal Reserve Bank would soon be out of business. The Christian Law Fellowship has exhaustively researched this matter through the Federal Register and Library of Congress. We can now safely conclude that this Executive Order has never been repealed, amended, or superseded by any subsequent Executive Order In simple terms it is still valid.

When President John Fitzgerald Kennedy - the author of *Profiles in Courage-* signed this Order. it returned to the federal government, specifically the Treasury Department the Constitutional power to create and issue currency - money – without going through the privately owned Federal Reserve Bank, President Kennedy's Executive Order 11110 [the full text is displayed further below gave the Treasury Department the explicit authority:

"To issue silver certificates against any silver bullion, silver, or standard silver dollars in the Treasury." This means that for every ounce of silver in the U.S. Treasury's vault the government could introduce new money into circulation based on the silver bullion physically held there. As a result more than $4 billion in United States Notes were brought into circulation in $2 and $5 denominations. $10 and $20 United States Notes were never circulated but were being printed by the Treasury Department when Kennedy was assassinated. It appears obvious that President Kennedy knew the Federal Reserve Notes being used as the *purported* legal currency were contrary to the Constitution of the United States of America. "United States Note? were issued as an interest-free and debt-free currency backed by silver reserves in the U.S. Treasury.

In the illustrations below, we compare a "Federal Reserve Note" issued from the *private* central bank of the United States (the Federal Reserve Bank A/K/A Federal Reserve System) with a 'United States Note" from the U.S. Treasury issued by President Kennedy's Executive Order. They almost look alike, except one says "Federal Reserve Note" on the top while the other says "United States Note". Also, the Federal Reserve Note has a green seal and serial number while the United States Note has a red seal and serial number.

FEDERAL RESERVE NOTE

UNITED STATES NOTE

President Kennedy was assassinated on November 22. 1963 and the United States Notes he had issued were immediately taken out of circulation. Federal Reserve Notes continued to serve as the legal currency of the nation. According to the United States Secret Service. 99% of all U.S. paper "currency" circulating in 1999 are Federal Reserve Notes.

Kennedy knew that if the silver-backed United States Notes were widely circulated, they would have eliminated the demand for Federal Reserve Notes. This is a very simple matter of economics. The USN was backed by silver and the FRN was not backed by anything of intrinsic value. Executive Order 11110 should have prevented the national debt from reaching its current level (virtually all of the nearly $9 trillion in federal debt has been created since 1963) if LB3 or any subsequent President were to enforce it. It would have almost immediately given the U.S. Government the ability to repay its debt without going to the *private* Federal Reserve Banks and being charged interest to create new "money" Executive Order 11110 gave the U.S.A the ability to. Once again, create its own money backed by silver and real value worth something. Again, according to our own research, just five months after Kennedy was assassinated, no more of the Series *1958* "Silver Certificates" were issued either, and they were subsequently removed from circulation. Perhaps the assassination of JFK was a warning to all future presidents not to interfere with the *private* Federal Reserve's control over the creation of money.

It seems very apparent that President Kennedy challenged the "powers that exist behind U.S and world finance". With true patriotic courage, JFK boldly faced the two most successful vehicles that have ever been used to drive up debt: 1) war (Vietnam): and, 2) the creation of money by a privately owned central bank. His efforts to have all U.S. troops out of Vietnam by 1965 combined with Executive Order 11110 would have destroyed the profits and control of the *private* Federal Reserve Bank.

The driver he turns around with a gun in his left and fires the fatal shot.

"Shot by his own driver" Footage from The Zepruder Film.

http://www.themeasuringsystemofthegods.com/Images/Animation.gif

View the video here and see for yourself. This New World Order will do anything to take back control of their agenda. Lee Oswald outside the same building at the same time

George Herbert Walker Bush was there. This man gave a New World Order speech do you think he had anything to do with it?

Then we have his son George Walker Bush when 9/11 happened. Even building 7 came down and it was never hit by an alleged plane. Something very strange there.

Concrete evidence of nuclear detonations at WTC on 9/11

Miniature thermal nuclear device. Proven 9-11 Nukes = US Government Involvement. By Dr. Ed Ward, MD 9-6-10 The Current Irrefutable Conclusion = Treason, Murder and Crimes Against Humanity for 9-11, the Iraq War, and Revocation of Constitutional Rights of US Citizens - All US 'representatives' appropriate US citizens, and non-citizens, need to be on trial in Constitutional Courts.

The Current 'Basic' Proven 9-11 Nuke Evidence - (About 95 to 98% of the TOTAL evidence covered) - All proven basic physics/chemistry ANY high school graduate should understand after the basic courses.

1. Three Massive WTC Craters - See us government LIDAR proof: - Nothing else known to man can leave ALL the WTC debris and this particular evidence in the length of time needed, except a third generation Micro Nuke - Mini Nuke - Nuke. It is 100% classic textbook nuclear event residue - ZERO ANOMALIES.
2.

2. Five Acres (1.2 Billion Pounds = Weight of Residue of 3 WTC Buildings (WTC 1, 2, and 6)) of WTC Land Brought to Seering Temperatures in a Few Hours by an 'Anaerobic, Chlorine Fueled "Fire" - Impossible by Basic Laws of Physics. See us Gov Thermal Images proof. http://letsrollforums.com/update-us-government-s-t22024.html?t=22024 - NOT an endorsement of the site.

3. Tritium Levels 55 Times (normal) Background Levels assessed a numerical value of 'traces' and 'of no human concern'. See us government (DOE report) proof: - Nothing but a NUCLEAR EVENT can cause 'tritium' formation - basic physics fact. http://groups.yahoo.com/group/EdWard-MD/message/141

4. An Impossible "Fire" (Combustion Process). See Laws of Physics for Fire/Combustion Process and Dr. Cahill's data on 'anaerobic incineration'.

http://www.rense.com/general77/newlaws.htm
5. 3 Billion pounds of building instantly turned into 2 Billion pounds of micronized dust.

http://www.serendipity.li/wot/ed_ward/use_of_abombs.htm
6. 16 inch steel Spires that withstood 1/2 a Billion pounds of building falling on them and suddenly turn into limp noodles and partially vaporize.

Image of Steel girders turned to dust.

7. Hiroshima effect cancers in responders and locals.
All of the above are facts are proven with referenced links of reputable data sources - many are from the government itself and more. Mini Nukes Were Used on 9-11
http://www.henrymakow.com/mini-nukes_were_used_on_9-11.html

Update: Mini Nukes at the WTC - 9-11 The Fetzer/Ward: Conversations on Mini Nukes in/on WTC/9-11 - 1 hour
http://radiofetzer.blogspot.com/2010/02/dr-ed-ward-md.html

Setting the Record Straight on Mini Nukes in the WTC/9-11
http://www.rense.com/general91/setting.html

Indeed, the nuclear component of the decimation of World Trade Centre buildings 1, 2, and 6–where WTC-7 appears to be a separate case–is the darkest and most closely guarded secret of 9/11. With so many folks claiming different theories, it is difficult for average people to know what to believe. Fortunately, we have scientific proof of what happened at Ground Zero. The dust and water samples reveal the true story of what happened on 9/11. This article thus provides more of the scientific evidence–especially from the USGS dust samples–that settles the debate in favour of the demolition of the WTC buildings as having been a nuclear event.

The Root: **Take a look at 'Operation Northwoods'**. The Basic Start a War by murder and treason - CIA Basic Operation Plan - 911/Cuba

Now jump in time to **Obama the Next "lap dog"**

Now we have Barry Soetoro AKA Barak Obama the U.S President. No Wonder he did not want to show his birth certificate. I always thought you have to be born in the USA to be president, not just be a Citizen. Do you see the power these people have with their black box voting system?

Who is in Control?

The Egyptian glyph for Mason is 'HUM-U' 'HUN' is the root word meaning "LIZARD" "HUM-U is literally "(one who is) of the Lizards"

> "We are grateful to The Washington Post, NY Times, Time Magazine and other great publications whose directors have attended our meetings and respected their promises of discretion for almost forty years. It would have been impossible for us to develop our plan for the world if we had been subject to the bright lights of publicity during those years. But, the work is now much more sophisticated and prepared to march towards a world government."
>
> - David Rockefeller
> address to a meeting of The Trilateral Commission
> June, 1991.

Quote: "No one will enter the New World Order... unless he or she will make a pledge to worship Lucifer. No one will enter the New Age unless he will take a Luciferian Initiation." David Spangler Director of Planetary Initiative United Nations. The above Quote: Is from a book called, 'Re-imagination of the World' by David Spangler and William Thompson. The NWO has control over just about everything and the people have let it happen. The Media, The false concepts within Science and Religions are all their work.

Der Erziehungsrat
des
Kantons Aargau

urkundet hiemit:

Herr <u>Albert Einstein</u> von <u>Ulm</u>, geboren den 14. März 1879,

besuchte die <u>aargauische Kantonsschule</u> & zwar die <u>III. & IV.</u> Klasse der <u>Gewerbeschule</u>.

Nach abgelegter schriftl. & mündl. <u>Maturitätsprüfung</u> am 18., 19. & 21. September, sowie am 30. September 1896 erhielt derselbe folgende Noten:

1. Deutsche Sprache und Litteratur	5
2. Französische "	3
3. Englische "	—
4. Italienische "	5
5. Geschichte	6
6. Geographie	4
7. Algebra	6
8. Geometrie (Planimetrie, Trigonometrie, Stereometrie & analytische Geometrie)	6
9. Darstellende Geometrie	6
10. Physik	6
11. Chemie	5
12. Naturgeschichte	5
13. Im Kunstzeichnen	4
14. Im technischen Zeichnen	4

*Hier gelten die Jahresleistungen.

Gestützt hierauf wird demselben das Zeugnis <u>der Reife</u> erteilt.

Aarau, den 3ten Oktober 1896.

Im Namen des Erziehungsrates,

Der Präsident:

Der Sekretär:

This a copy of Albert Einstein's school report. If you want to know a 6 is the worst grade one can have and 1 is the best. This man was a fraud and plagiarist and chosen by the JEWS as a front man for a NEW AGE of Pseudo- physics that has stagnated the development of humanity. **Shame on the lot of you.** There is so much information about him being a fraud I am surprised they still teach that rubbish. People, please research this.

Nikola Tesla should have got all the recognition Einstein's Idiots – YouTube
www.youtube.com/playlist?p=PL70689492C45363CD

In Closing

If you want to make a difference in this world, you are going to have stand up for your rights before they are all gone. A whirlwind of information surrounds you so take the time to step back and see what is going on. Listen to your heart and devote your mind to achieving Krsna Consciousness.

Let there be ONE god for the whole world Sri Krsna and one mantra only- Hare Krsna, Hare Krsna, Krsna, Krsna, Hare Hare, Hare Rama, Hare Rama, Rama, Rama, Hare Hare.

It is my will and pleasure to be in the service of the Supreme personality, The ONE.

"Everything is nothing, nothing is everything."

Bibliography

"Bhagavad-Gita (AS IT IS)" By A.C Bhaktivedanta Swami Prabhupada 2008
"The Secret of Light" By Walter Russell 1947
"The Universal ONE" By Walter Russell 1926
"In The Wave Lies the Secret of Creation" Walter Russell. By Dr Timothy Binder 2006
"Harmonic Proportion and Form in Nature, Art and Architecture" By Samuel Colman 1912 *****
"The Power of Limits, Proportional Harmonies in Nature, Art and Architecture" Gyorgy Doczi 2005 *****
"Mysterium Cosmographicum The secret of the Universe" By Johannes Kepler (1596) 1981
"The Harmony of the World" By Johannes Kepler (1599) 1997
"The Three Books of Occult Philosophy" By Henry Cornelius Agrippa 2009
"THE MAGUS, OR CELESTIAL INTELLIGENCER" BY FRANCIS BARRETT 1801
"The Ultimate Reality" By Joseph H. Cater 1998 same as "The Awesome Life Force" by Joseph Cater 1984.
"The Dimensions of Paradise" By John Michell 1971
"City of Revelation" By John Michell 1972
"How the World is Made, The story of creation according to Sacred Geometry" By John Michell and Allan Brown
"A Beginner's Guide to Constructing the Universe" By Michael s. Schneider 1994
"Music of the Spheres Volume 1 and 2" By Guy Murchie 1967
"Mysteries of Magic: A Digest of the Writings" By Eliphas Levi 1993
"Occult Chemistry" By C.W Leadbeater and Annie Besant 1919
"The Etheric Formative Forces in Cosmos, Earth and Man" Dr Guenther Wachsmuth 1932
"New Theories of Matter and of Force" By William Barlow 1885
"A Sketch of a Philosophy Part 2 MATTER and Molecular Morphology" By John G Macvicar 1874
"Implosion, the Path of Natural Energy" by Viktor Schauberger 1985
"Universal Laws Never Before Revealed, Keely's Secrets" By Dale Pond 2004
"The Handbook of Unusual Energies" By J.G Gallimore 1974
"The Secret Doctrine" By H. P. Blavatsky 1875
"THE PHYSICS OF THE SECRET DOCTRINE" BY WILLIAM KINGSLAND 1910
"THEORY AND CALCULATION OF ALTERNATING CURRENT PHENOMENA" BY CHARLES PROTEUS.STEINMETZ 1900
"VRIL or Vital Magnetism, Secret Doctrine of Ancient Atlantis, Egypt, Chaldea and Greece" McCLURG & CO 1911
"Technology of the Gods, the Incredible Sciences of the Ancients" By David Hatcher Childress 2000

"THE MASTER MASON" Authorized by GRAND LODGE F. & A. M. Compiled by COMMITTEE ON MASONIC EDUCATION
"Magnetic Current" By Ed Leedskalnin 1945
"Graeco- Egyptian Work upon Magic, From a Papyrus in the British Museum" By Charles Wycliffe Goodwin 1852
"John Dee (1527-1608)" By Charlotte Fell Smith 1909
"The Enochian Evocation of Dr John Dee" By Geoffrey James
"The Complete Enochian Handbook" By Vincent Bridges 1996
"LUCIFER a Theosophical magazine" By H.P Blavatsky and Mabel Collins September 1887 – February 1888
"The Sacred Theory of the Earth" Dr Thomas Burnet 1699
"A Primer of Higher Space" By Claude Bragdon 1913
"ELECTRICITY AND MATTER" By J.J Thomson 1904
"Reich of the Black Sun" By Joseph P. Farrell
"Corporate UNITED STATES 3rd Edition" By Sir David Andrew 2008
"THE TEMPLES OF BAAL" By SIR DAVID ANDREW 2007
"The Terra Papers" By Robert Morning Sky 1996
"THE MANU: An Interdimensional Artifact at the Genesis of History" By Ananda and ATON VASE DA 2005
"Scalar Waves (first Tesla physics text book)" Prof. Dr. Konstantin. Meyl, 1996
"Occult Ether Physics" By William Lyne
"Pentagon Aliens" By William Lyne
"Chariots of the Gods, Unsolved Mysteries of the Past" By Erich von Däniken 1968
"The Vaimānika Shāstra वैमानिक शास्त्र ("Science of Aeronautics") G.R Josyer 1973
"Vimanas Ancient Flying Saucers of Atlantis and Lemuria" By Mary Sutherland 2010
"The Rational Non-Mystical Cosmos; the Mysticism of Science Exploded." Gillette, George Francis. 1933.
"TOP SECRET BANKER'S MANUAL. FOR BANKERS ONLY" Tom Schauf
"Zetetic Astronomy" By Lady Blount & Albert Smith. 1956.
"The Popularity of Error and the Unpopularity of Truth" BY JOHN HAMPDEN, ESQ 1869
"The Plane Truth" By Samuel Shenton 1966
"THE HOLLOW GLOBE OR THE WORLD'S AGITATOR and RECONCILER. A TREATISE ON THE PHYSICAL CONFORMATION of THE EARTH". Presented by M. L. SHERMAN, M. D And "Written by PROF. WM. F. LYON. 1871.
"A History of the Theories of Aether and Electricity" Edmund Whittaker 1910
http://theflatearthsociety.org/cms/
VIDEOS
Holographic Universe, The Secret beyond Matter
https://www.youtube.com/watch?v=UDnakYnA7VA

Arthur C Clarke - Fractals - The Colors of Infinity
https://www.youtube.com/watch?v=Lk6QU94xAb8
Vedic Tour of our Universe and Beyond
https://www.youtube.com/watch?v=mKyQxI1Q3UU
Scientific Verification of Vedic Knowledge *Full*
https://www.youtube.com/watch?v=uKyn3FDaTYc
Vimanas- ancient flying machines of India
https://www.youtube.com/watch?v=JKbQiKrqaIw
Documentary Unlocking the Mystery of Life Intelligent Design
https://www.youtube.com/watch?v=tzj8iXiVDT8

www.ingramcontent.com/pod-product-compliance
Lightning Source LLC
Chambersburg PA
CBHW071354170526
45165CB00001B/44